SATELLITE TELEVISION IN WESTERN EUROPE

Revised edition

Richard Collins is Senior Research Associate, Department of Media and Communications, Goldsmiths' College, University of London

SATELLITE TELEVISION IN WESTERN EUROPE

Revised edition

Richard Collins

Acamedia Research Monograph: 1

British Library Cataloguing in Publication Data

Collins, Richard
   Satellite Television in Western Europe. –
   2 Rev. edn. – (Acamedia Research Monograph series)
   I. Title   II. Series

ISBN: 0 86196 388 1
ISSN: 0956-9057

Series Editor: Manuel Alvarado

Published by

John Libbey & Company Ltd, 13 Smiths Yard, Summerley Street,
London SW18 4HR, England
Telephone: 081-947 2777 – Fax: 081-947 2664
John Libbey Eurotext Ltd, 6 rue Blanche, 92120 Montrouge, France
John Libbey - C.I.C. s.r.l., via Lazzaro Spallanzani 11, 00161 Rome, Italy

© 1992 Richard Collins. All rights reserved.
Unauthorised duplication contravenes applicable laws.
Printed in Great Britain by Whitstable Litho Ltd, Whitstable, Kent, U.K.

Contents

Preface to the revised edition		1
Introduction		3
Chapter 1	The history of satellite television	13
Chapter 2	The shaping of satellite television	17
Chapter 3	The funding of satellite television	43
Chapter 4	The audience	55
Chapter 5	European television satellites	69
Chapter 6	The future of satellite television in Europe	93
Chapter 7	The lessons from the second generation of satellite television	105
Conclusion		121
References		123

Preface to the revised edition

Three years after completion of the typescript that became the first edition of this study the conclusions I advanced require qualification but not wholesale revision. In mid 1989 I wrote 'The prognosis for satellite television in Europe is rather gloomy' and argued that satellite television would succeed (or fail) in so far as it competed successfully with terrestrial television services. That remains broadly true. Where European satellite television has been most successful is where terrestrial television has been relatively weak, for example in small countries where terrestrial services have been relatively poorly funded and in countries where the hold of terrestrial services on the attention of viewers has been poor because of underprovision of entertainment: Germany is the best example of the latter.

In the UK satellite television is in about 10 per cent of UK television households where it accounts for about 30 per cent of viewing. Thus the four terrestrial networks still account for the overwhelming majority of UK television viewing. Though it's clear that a significant number of households value sufficiently the extension of viewing choice which satellite television provides to defray the costs (considerable in comparison to the costs of watching terrestrial television) of consumption of satellite services. One factor which has changed since I wrote the first edition of this book is that in both real and nominal terms the cost of reception of satellite television has declined. In early 1992 the price of receiving hardware for BSkyB services was £180 (installation cost an additional sum, from £70). The subscription to BSkyB film channels was £12 per month for one channel and £17 per month for two. Thus the annual cost (in interest forgone on capital, with a ten year depreciation of equipment and direct subscription costs) of receiving BSkyB satellite television in the UK in 1992 ranged from *ca* £50 (with no film channel) to *ca* £254 per annum (with both channels). This is a small incremental cost beyond those of receiving terrestrial television. Moreover the cost of the licence fee required to receive terrestrial television rose between 1989 and 1992, from £62 to £77 pa.

In spite of the effective bankruptcy of one of two UK Direct-to-Home Satellite television services, its merger with its former rival and the continuing financial and operating deficit experienced by the new, merged, service it is clear that satellite television is here to stay. But it remains to be seen whether satellite television will become a major element in UK viewers' consumption of television. However it is clear that consumption has risen beyond the 'derisory level' which I noted in 1989 and that satellite television may occupy a place in the UK television market similar to that which the independent/syndication broadcasting sector occupies in the United States rather than simply failing as I then thought more likely. But it should not be forgotten that one lavishly funded UK DBS service – BSB – went belly up and the surviving service – BSkyB – has yet to consistently return a trading profit still less pay down its sunk costs (see Conclusion for the history of BSB, Sky television and BSkyB).

My 1989 discussion of the prospects for satellite delivery of erotic television also demands revision in the light of subsequent events. Promulgation of the 'Television without Frontiers' Directive by the European Community has meant that UK regulators have lacked power (since October 1991 when the Directive took effect) to prohibit cable retransmission of satellite television from other European Community states. Thus the RTL Plus signal, which was not formerly retransmitted in its original broadcast form by UK cable networks because of its erotic content, must now be retransmitted with the jiggles and nipples to which German viewers have long been accustomed. Moreover it has proven to be considerably easier, than I foresaw in 1989, for UK viewers to acquire unauthorized Filmnet decoders and thus secure access to Filmnet's erotic programming.

The changes made in this, revised, edition are of two kinds. I have made some minor editorial changes to the original text (correcting typographical errors, rectifying minor inconsistencies and making occasional changes to the tenses of verbs and to adverbs of time and updating information, though I have not replaced original usages such as 'Soviet Union' and 'West Germany' by others). And I have added a new chapter to the original text which discusses the development of Direct-to-home Broadcasting of television by Satellite (DBS) in the UK and thus updates the discussion of DBS which I put forward in the original edition's final chapter (now the penultimate chapter but still numbered Chapter 6).

The new final chapter first appeared as PICT Policy Research Paper No 15 which was commissioned and first published by the Economic and Social Research Council's (ESRC) Programme on Information and Communication Technologies (PICT). Between the publication of PICT Policy Research Paper 15 in 1991 and writing this new introduction (in early 1992) the UK DBS service BSkyB had extended its reach into an estimated 2.5 million UK television homes. It has added a Comedy Channel to its bundle of services (Sky 1, Sky News, Sky Sports, Movie Channel, Sky Movies, Comedy Channel) and the Sky services accounted for 22.3 per cent of viewing in UK satellite and cable homes (other non-terrestrial cable and satellite channels accounted for 7.7 per cent of viewing in cable and satellite homes).

At the time of writing BSkyB had not achieved a trading profit though its early move into a trading profit (*ie* before finance charges) was anticipated. In 1991 BSkyB was estimated to have an income of £50 million pa from advertising and £240 million pa from subscriptions to its film channels. An estimated 70 per cent of BSkyB homes subscribe to movie channels (the churn rate[1] is put at 20 per cent). BSkyB's costs were estimated to be *ca* £280 million pa.

Amortisation of the very substantial costs sunk in the channel will take much longer. BSkyB's owners are likely to have to wait until the next millenium for a return on their investments – particularly if competitors enter the satellite television market as Thames Television (the long established London terrestrial television broadcaster which lost its franchise in the 1991 UK terrestrial television franchise renewals) has promised to do. Who knows what television will look like in the next millenium?

Richard Collins
London, November 1992

---

1 'Churn' signifies the proportion of subscribers to pay television who do not renew their subscriptions.

Introduction

It is a truism to state that communications are undergoing rapid and comprehensive change. New technologies and new ideologies are re-articulating the relations between old-established communication media.

For long the distinction between media of carriage (such as the postal and telephone systems) and media of content (such as the newspaper press and broadcasting) held good. A related, but not perfectly congruent, distinction between point-to-point communication, such as telephony, and point-to-multipoint communication, such as television, also prevailed. Such old paradigms have now been displaced by new notions of 'convergence' between previously distinct media, and a generic notion that we are in, and the media are part of, a 'wired society'. Convergence is evident between telecommunications and broadcasting, for example in cable television, and between telecommunications and print, for example in videotex and fax. Point-to-multipoint 'broadcast' messaging, which used to be a quintessentially 'public' form of communication, is now used for addressing selected and exclusive receiver populations through such media as datacasting (the transmission of messages accessible only to designated receivers, such as the BBC's use of its 'night hours' for transmission of encoded information to the video cassette recorders of medical practitioners). And point-to-point communication via the public switched telephone network, (PSTN), hitherto 'private' communication, is now used in point-to-multipoint modes (as in electronic mail).

However these new paradigms are also under pressure and are being revised in what seems to be a permanent revolution in communications infrastructures, and consequently in the social relations which they shape and by which they are, reciprocally, shaped. The wired society promises to be replaced – even before one can be said to have been created – by a wireless society. As higher and higher frequencies in the radio spectrum are utilized, following improvements in radio transmission and reception technologies, more and more spectrum bandwidth becomes available for wireless messaging, both in point-to-point and point-to-multipoint modes. Thus wireless communication substitutes for systems that were hitherto wired.

To designate wired and wireless communication as antitheses is, of course, an over simplification. They co-exist now, as in the past, in complementary as well as in contradictory relationships. What is distinctive about the present is the flux in, and rapidity of, the processes of change (which make a study such as this subject to continual revision).

Formerly, the balance between wired and wireless communication shifted, but in fairly long temporal waves. Wireless semaphore signalling was replaced by the wired telegraph; decades passed before the telegraph was supplemented by wireless radio communication. Long distance terrestrial radio communication (at the frequencies then used) did not have either the transmission capacity to carry a growing volume of traffic or the reliability of wired links, and so wired communication, such as submarine and terrestrial telecommunications cable links, remained and grew at the same time as did wireless messaging. However research and development in the microwave frequency range (1 to 30 GHz) later delivered economies that made microwave links for long haul telecommunications traffic cheaper than fixed links, thereby shifting the balance towards wireless systems. Further recent changes in cost/benefit relationships between telecommunication technologies are leading to a return to wired, fibre-optic, transmission systems rather than further development and exploitation of wireless microwave communication. Neither wired nor wireless systems have ever decisively supplanted the other, though at different moments in communication history there have been definite emphases which have constituted historical periods as ones of wired or wireless dominance.

Characteristically telecommunication (including television) message paths consist of both wired and wireless elements. In the present instance the speed of innovation and change is very high, and new communication technologies are shifting the balance between wired and wireless systems very rapidly. This balance depends on a variety of factors, of which the technology/cost factor is particularly important. It is one which has changed very rapidly in the last decade under the influence of the plethora of new technologies including the technology of communication satellites. Such satellites, which made possible unprecedented exploitation of Super High Frequency sections of the radio spectrum for long distance communication, were once thought to have made wired long distance communications obsolete. However advances in fibre-optic technologies have made very long-haul wired communication, *eg* trans ocean submarine cable links, competitive with wireless satellite communication. The relation of wired and wireless transmission systems thus remains in continual flux.

The uncertainty, and opportunities, created by technological change are amplified by changes in the 'culture' and practice of communications regulation. In the UK the regulatory changes are exemplified by the establishment of new regulatory bodies (Oftel, the Cable Authority, the Broadcasting Standards Council, the Radio Authority) and changes to existing regulators, notably the metamorphosis of the Independent Broadcasting Authority into the Independent Television Commission (incorporating the Cable Authority). Changes such as these signify a deep uncertainty about the role, procedures and legitimacy of regulation in the contemporary communications order. On one hand the Government enjoins, and facilitates, the 'withering away' of regulation, judging it to be unnecessary when effective competition can be established. On the other hand creation of competition seems to necessitate an increasingly byzantine apparatus of regulation in order to ensure that established enterprises are unable to use their market power to crush new entrants and exploit final consumers.

Within this context of expanding capacity for electronic communication in wired or wireless modes, both point-to-point and point-to-multipoint; of uncertainty as to whether or not the supply of information (*eg* television programmes) will expand to fill the new capacity; of convergence between hitherto separate technologies and services; and of rapid technological and regulatory change, the future role of satellite television in Western Europe is hard to analyse with confidence (not least because of the unavailability and unreliability of essential data). Nevertheless satellite television demands attention as a phenomenon which is set to challenge long-established assumptions about the role of broadcasting, the nature of national cultures, the linkage between political systems and cultural consumption, and the relation between public and private provision of broadcasting services. Everywhere there is a consensus that satellite television is important; but nowhere can it be specified with any confidence what that importance is likely to be.

Satellite television has attracted attention for two reasons. First it abolishes the traditional relationship between cost and distance of transmission, and potentially offers new market stratifications. If signals are propagated over a wide geographical area, for example over Western Europe, an integrated advertising market and audience of an unprecedented size and structure is potentially available. Such a potentiality is attractive to interests such as the European Commission, charged with the creation of a single EEC market by 1993, and threatening to those who, like Jack Lang, Minister of Culture and Communication in France, fear the erosion of existing communities and cultures. Lang referred to television satellites as 'Coca cola satellites attacking our artistic and cultural integrity' (cited in *Financial Times* 30.4.1984 p. 3).

Second, satellite television, by using Super High Frequencies (SHF) hitherto without practical usefulness for television broadcasting, promises to abolish radio spectrum scarcity which has been thought to constrain development of additional television services to compete with the terrestrially broadcast channels established in West European states. Satellite television, therefore, offers hope to those wishing to introduce more competition and more choice in television and a threat to those who wish to retain the status quo.

The potentiality of satellite television to transform television markets undoubtedly exists. But there are many difficulties in the way. The differences of language (and to a lesser extent culture) in Western Europe have slowed consumption of trans-national television and the establishment of a trans-national audience. The paucity of trans-national brands has inhibited development of an integrated European television advertising market. Frequency planning authorities are proving successful in 'finding' additional frequencies for terrestrial broadcasting which promise to expand the supply of television, and increase competition, at lower cost than will satellite television.

The most important factor that has shaped the history of satellite television in Western Europe and which will finally determine its success, or failure, is its capability of delivering to viewers TV programming which has equal, or preferably superior, benefits in relation to costs, compared to competing terrestrial services. Since terrestrial television differs in different locations in Western Europe it follows that satellite television will enjoy greater, or lesser, chances of financial success in different European markets depending, among other things, on the strength of the competition offered by terrestrial television. However there are a number of common factors that apply in different local television markets but which, depending on their different relative importances in different markets, produce different outcomes. Such factors are discussed extensively in the body of this study but the most salient are briefly introduced below.

Costs

In the United Kingdom the existing four UK broadcast television channels provide information, education and entertainment to final consumers at very low cost: estimated (Ehrenberg and Barwise, 1982) at 1.5 p per hour for ITV and 2 p per hour for BBC. To receive satellite television final consumers have to either subscribe to a cable network or purchase a TVRO (Television Receive Only Earth Station), otherwise known as a 'dish'. Viewers will also need to purchase or rent additional dishes and receivers if they wish to receive signals from more than one satellite and one or more decoders if they wish to receive scrambled signals, such as feature film channels available only on a pay basis. The cost of reception of satellite television is therefore considerably in excess of the, very low, cost of consumption of terrestrial television. If further terrestrial broadcast channels are established in the UK then the existing poor outlook for satellite television there is likely to worsen. The UK is perhaps the most striking West European instance of a general rule: unless satellite television can deliver programming more attractive (which in effect means more costly) to final consumers than that available via competing distribution media, then uptake of satellite services is likely to be low.

The audience

Terrestrial television is cheap in relation to competing systems, but the audience that it delivers to advertisers and programmers is largely national. The wide 'footprints' of satellite television offer advertisers and programmers opportunities to restratify European audiences in new ways. Satellite television programmers have seen opportunities to agglomerate audiences across Europe so as to create a sufficiently large and prosperous audience for new programme services, such as an Arts channel, that are uneconomic on a national basis. France has led development of a high culture channel 'La Sept' in the hope that a Europe-wide audience will sustain a service which is uneconomic within a solely national context. The earlier and unsuccessful 'Europa' initiative led by the Dutch broadcasting authority NOS attempted something of the same.

Advertisers too may be interested in a television advertising medium with Europe-wide reach which gives them access to either the European mass audience or a particular audience segment (for example 'Yuppies' or international business financial and political élites) that can be addressed only imperfectly and wastefully by established mass national television services.

To these potential 'horizontal' groups in the European public should be added potential 'vertical' groups of audiences dissatisfied with existing national television services. But whether programming is directed to horizontal or vertical audience groups it has to be paid for.

Finance

Satellite television has proved attractive to advertisers wishing to reach audiences in markets where terrestrial television advertising time is undersupplied. This has happened most importantly in West Germany where terrestrial broadcasting has provided a poor service to advertisers. There the satellite channels RTL Plus and Sat Eins deliver more entertaining programming than their terrestrial public service competitors and thus successfully attract and retain audiences. Moreover satellite television in Germany is much more advertiser-friendly than the terrestrial ZDF and ARD networks.

Advertising finance potentially offers a means of funding satellite television, but the experience of European satellite television to date suggests that:

(1) No trans-national advertising market exists.

(2) There is advertiser-resistance to funding certain types of programming even though it may be popular with final consumers.

(3) Even in large and prosperous markets undersupplied with television advertising and popular programming, sufficient advertising revenues are not yet available to cover costs, still less to yield profits for satellite television.

The principal alternative to advertising revenue as a funding mechanism is subscription funding, but this system involves significant transaction costs. Subscription channels require, therefore, to deliver to final consumers programmes offering benefits which outweigh not only the costs of programming equal or superior to that available through competing distribution media (notably terrestrial television) but these transactions cost as well. Here, too, the experience of satellite television services in Europe has not been particularly positive.

Programme services

Given the higher costs of satellite television *vis-à-vis* terrestrial television it must deliver programming superior to terrestrial services if it is to be successful. In this area, one advantage of satellite television is that it has the potential to deliver High Definition Television (HDTV) more effectively and economically than can terrestrial broadcasting. HDTV requires a greater bandwidth than conventional services, for which there is unlikely to be space in the frequency range used for terrestrial broadcasting. If, as some predictions suggest, final consumers do desire HDTV they may be willing to pay additional reception and transaction costs in order to secure it and thus offer market opportunities to satellite television suppliers. HDTV thus offers a possible scenario for successful development of satellite television services. However HDTV has not yet progressed beyond the development stage in Europe.

In default of satellite TV offering picture quality of a higher *technical* standard than terrestrial services, the success of satellite television will be conditional on its supply of programming being either unavailable to viewers through other distribution means, or superior to that available at equivalent costs by other means.

Certain categories of television programming are terrestrially undersupplied, or censored, for reasons of public policy. Programme content such as pornography, violence and material affecting the security of the state are generally unavailable to viewers even though they may wish to consume such material. However, satellite television is unlikely to supply programming in any of these fields. Advertisers have shown themselves unwilling to associate their products with programming that they regard as offensive, even though such programming attracts substantial audiences. Tesco Stores' withdrawal of advertising from the UK daily newspaper the *Daily Star*, and Sky Channel's difficulties in securing advertising for its highest rated programme *Wrestlemania* (perceived by advertisers as too 'down market') suggests that advertising finance will not be forthcoming for violent, or pornographic, programming.

The only possible alternative to advertising finance, for such programming, is subscription. To collect subscription revenue, programme providers need to control access to programming. This they can do either via cable distribution and/or by scrambling transmissions and selling or renting decoders to viewers. Both cable distribution and the sale or rental of decoders are susceptible to effective control by government. It is unlikely, therefore, that satellite television will provide the programming that public policy prohibits on terrestrial television.

For satellite television to succeed in competition with terrestrial television it must provide more attractive programming to viewers than does its terrestrial competitor. However terrestrial broadcast television is always significantly advantaged in competition for audiences with satellite TV. Its distribution costs are low, it has been long established and it consequently enjoys a dominant market position. In many cases terrestrial television has the significant competitive advantage of transmitting advertising-free programming. And, most important, in large European countries broadcasters have enjoyed funding sufficient to finance a range of high cost programmes. The programme offer of satellite television is, therefore, all other things being equal, likely to be cheaper and less attractive than that of terrestrial television.

However there are two West European environments in which satellite television has enjoyed some success: the small countries of Western Europe, and West Germany.

Small countries offer promising opportunities to satellite television because national terrestrial services do not dispose of sufficient revenues to fund high budget national programming in sufficiently large quantities to offer viewers a range of choices. Satellite television may potentially, therefore, offer to viewers a welcome extension of choice and/or improvement in programme cost and quality. This was once the case in the Netherlands, but between 1985 and 1987 viewing of Sky Channel and other satellite channels *declined*. A survey of cable subscribers in Amsterdam showed that there was more interest in receiving foreign *terrestrial* television signals than satellite services.

Moreover small countries which potentially offer opportunities to satellite television can, and do, take steps to close their windows of vulnerability. The recent restructuring and rescheduling of Dutch public television's programming effectively regained some of the audiences that had been lost to satellite television. As Wim Bekkers (Head of Audience Research NOS Netherlands) stated: 'The most important challenge for public broadcasting in the commercial satellite era is to compete and survive with entertaining, educational, cultural and news programmes'.

West Germany provides an exception to the general rule that satellite television, particularly in large states, has had little impact on viewing habits or advertising markets. The success of the two commercial satellite channels transmitted to West Germany, RTL Plus and Sat Eins, is due to the distinctive character of terrestrial broadcasting in West Germany.

The three channels of public service terrestrial broadcasting in West Germany offer a poor medium for advertisers. Advertising is confined to 'blocks' separated from programmes and transmitted between 6 and 8 pm (and not on Sundays or religious holidays). It is sold by an organization at 'arm's length' to programmers so advertisers are unable to locate advertisements next to particular desired programmes. And programming on West German terrestrial television is not designed to maximize ratings and exposure of audiences to advertisements. In contrast satellite television offers advertisers an attractive package of popular programming and twelve minutes of spot advertising per hour with an audience-maximizing programming orientation.

West German terrestrial public television has long had the character of a secular church and the success of RTL Plus and Sat Eins in attracting advertising revenue and a growing proportion of audience attention testifies to the vulnerability to commercial competition of public sector broadcasters that have lost contact with popular taste. But even in West Germany, though close to economic viability, satellite television still accounts for only a small proportion of total viewing.

Audience response to UK terrestrial television is difficult to evaluate. The Peacock Report (Peacock, 1986) demonstrated that there is an imperfect market in broadcasting services, that service providers get poor, and possibly misleading signals from consumers, and that audience behaviour is not necessarily indicative of audience wants. High consumption of existing programming is indicative only of preferences between existing alternatives and not necessarily of an optimal matching of supply to either actual or latent demand. Peacock pertinently cited the findings of the National Consumer Council that 'it would not be wise for broadcasters to assume that consumers think that everything is wonderful in the world of British broadcasting', and that the 46 per cent of television viewers who expressed satisfaction with UK television was 'a very low figure'. Even so, there has been little enthusiasm for either cable or satellite television in the UK. Consumption remains at a derisory level. The cost/benefit analyses performed by viewers have resoundingly favoured the existing terrestrial services.

International comparisons suggest four reasons for the success of non-terrestrial broadcast television services (such as cable and satellite), none of which currently apply to the UK:

(1) Access to television services relayed from a neighbouring country with a comprehensible language and better-funded television. The highest penetration of cable is in countries such as Belgium, the Netherlands, Canada and Switzerland. In these countries cable delivers, at low cost, access to well-funded terrestrial broadcast television from neighboring countries in a comprehensible language.

(2) Under-delivery of entertaining programming by national terrestrial broadcasters. The BBC's past loss of audiences to Radios Luxembourg and Normandy in the 1930s, to ITV in the 1950s, and to pirate radio in the 1960s and 1980s, was due principally to its patrician, 'improving' and boring programming. The loss of

Canadian viewers to American television has been, in some part, for the same reason: even francophone Canadians, partially insulated by language from the attractions of American television, adversely compare French Canadian television – 'une télévision vieillissante' – to American television – 'une télévision vehiculante la richesse, le rêve et l'éspoir'.
(Canada/Quebec, 1985, p. 47)

(3) Under-delivery of a satisfactory terrestrial medium for television advertising. It's not just a question of time being made available. US television has twelve (sometimes more) minutes of advertising each hour. These advertisements are intrusive and are not (as UK advertisements are supposed to be) at natural breaks in programmes. In such circumstances audiences may avoid advertising-financed television, even if the advertising-free alternative, such as cable, is available only at higher cost.

(4) Under-delivery of a technically satisfactory sound and image quality. This is one of the reasons why cable television (and direct-to-home reception of satellite television) in the USA is attractive to final consumers. Many US television viewers are unable to receive broadcast signals of a quality comparable to that taken for granted in Western Europe. Cable television in the US began as a medium which would rectify the technical deficiencies in broadcast signal quality. The introduction of HDTV as a non-terrestrial broadcast service may offer a similar scenario in the UK.

It is clear however that success or failure of satellite television depends primarily on its programming offer. Here satellite television is at a disadvantage. For satellite television revenues are low in comparison to those of terrestrial broadcasters. Recent estimates suggest that:

(1) Satellite TV will be able to make little programming and then only in low cost genres.

(2) Most satellite TV programming will be acquired on the international programme market or 'cascaded' from the archives of owners.

DBS

Satellite television is not a completely unitary phenomenon. In 1988 a technology, DBS (Direct Broadcasting by Satellite), which had been introduced earlier in Japan, entered the European stage.

Powerful television satellites (DBS) reduce the costs of reception incurred by viewers, thus improving the relative position, in cost/benefit terms, of satellite television *vis-à-vis* terrestrial television. But this decline in costs may not be sufficient to render satellite television sufficiently attractive to viewers where terrestrial television already delivers acceptable, if not optimal, programming.

Conclusions

In present European conditions satellite television is unlikely to be a serious presence in the UK, and most other West European broadcasting markets. Where satellite television is currently successful (though nowhere generating revenues sufficient to cover operators' costs) it remains very vulnerable to changed policies and practices of terrestrial broadcasters.

The existence in Western European television markets of a plurality of satellite television services testifies more to the lobbying power of the European aerospace and electronics industries than to viewer demand for new television services.

The key to the future of satellite television is the terrestrial television regime. In the UK, if the present terrestrial services are maintained (still more so if expanded) satellite television is unlikely to make a serious impact on UK viewing.

Accordingly a future of intense competition between television broadcasters can be anticipated, in which those with lowest costs relative to revenues and high capitalization will survive. The cost structure of terrestrial broadcasters will favour them in this competition. Their distribution systems are cheaper than those of satellite (and cable) operators and they enjoy large stocks of programming, the costs of which have already been written off. Satellite programmers will increasingly be forced to programme their services with programmes expensively acquired 'off the shelf'.

It is possible that the television market in the UK may therefore evolve in a manner similar to that of the television market in the USA. Well-funded broadcasters (in the USA Home Box Office, drawing its revenues from subscription, and the ABC, CBS and NBC advertising-financed networks) produce original programming, while the syndication/independent television stratum programme repeats. But the US market is bigger and richer than that of the UK and the cost structure of terrestrial broadcasting is more favourable than that of satellite television. It therefore seems more likely that satellite television in the UK will simply fail rather than assume a position equivalent to that of the, profitable, syndication/independent broadcasters in the USA. The experience of the UK is unlikely to be very different to that of other European states. The prognosis for satellite television in Europe is therefore rather gloomy.

1 The history of satellite television

In October 1945 *Wireless World* published an article by Arthur C. Clarke (now better known as a science fiction writer and most recently as Chancellor of the University of Moratuwa, Sri Lanka), which theoretically demonstrated the possibility of establishing a communication satellite in a geostationary orbit above the earth. (A geostationary, or geosyncronous, orbit is one in which a satellite maintains a constant position relative to the earth.)

An 'artificial satellite' at the correct distance from the earth would make one revolution every 24 hours: *ie* it would remain stationary above the same spot and would be within optical range of nearly half the Earth's surface. Three repeater stations, 120 degrees apart in the correct orbit would give television and microwave coverage to the entire planet.

A satellite's position relative to the earth is determined by the height, velocity and location of its orbit. Clarke reasoned that a satellite 'parked' 35,786 kilometres above the Equator would orbit the earth so that its position relative to the earth's surface would remain constant, or 'geostationary'. ('Parking' is a misleading metaphor because a satellite will only maintain such a geostationary orbit if it maintains a velocity of 3075 metres per second.) Powerful rockets, capable of locating satellites in a Clarke orbit, and miniaturization of electronic components, have made possible practical realization of what Clarke envisaged only as a theoretical possibility. The geostationary, or Clarke orbit, though making possible continuous provision of communication services (whether television or telecommunication) with relatively cheap ground stations, is by no means the only orbit used for communication or television satellites. And television satellites are but one, comparatively recent, instance of a general class of communication satellites.

The first satellite to be established in a geostationary orbit was Early Bird in 1965. Early Bird, launched by the international telecommunications satellite agency Intelsat, transmitted signals across the Atlantic interconnecting West European and North American telecommunications networks. It had a modest capacity of 240 voice telephony circuits, or one television channel. Early Bird was an example of a class of satellites known as fixed service, or telecommunications, satellites. Such satellites were not designed for the transmission of television signals for reception by final consumers, but a number of factors – including development of gross surplus capacity – have led to fixed service satellites being used for television broadcasting. Early Bird was an early instance of the appropriation of satellite systems, designed for point-to-point telecommunication traffic, and for point-to-multipoint television transmission.

Even though its capacity was equivalent to only one TV channel, Early Bird was not quite the first television satellite. From the 1950s satellites in non-geostationary orbits were used to relay television signals. These satellites required large, complicated and very expensive earth stations in order to track their movement across the sky. For example the antenna of AT&T's earth station at Andover, Maine, used to track satellites in non-geosyncronous orbits during the few minutes they were 'visible' to the earth station, weighed 380 tons. The UK equivalent, built at Goonhilly, Cornwall, has a 26 metre diameter antenna.

AT&T's Telstar, which relayed television across the Atlantic from 1962, is the best known of these first, non-geosyncronous, television satellites. It orbited the earth every 157 minutes and was capable of relaying signals across the Atlantic for 18 minutes of each orbit. On 23 July 1962 Telstar was used for the first live transatlantic television programme exchange. After transmitting a pot-pourri of images from the USA to Europe, on its next orbit, two and a half hours later, Telstar transmitted a reverse flow of images of the UK, France and West Germany to North America. As the Telstar example shows, satellites in non-geosyncronous orbits are unable to provide 24-hour relay services. Further, they require up and down linking of signals, to and from earth, to be performed by expensive steerable antennae at dedicated earth stations.

Satellites in geosyncronous orbit have none of these problems and from the mid-sixties have been used by broadcasters to improve the quality and immediacy of their coverage of world events, events, particularly of news and sport. Only polar regions are outside their range, so the Soviet Union uses four satellites in successive elliptical orbits to serve its Northern latitudes.

In general, the 1960s and 1970s saw Clarke-orbit satellites established everywhere as important media for the distribution of television signals. The satellites of this period were small and, by contemporary standards, low powered. Their antennae did not need to be steerable, but they were nonetheless comparatively large and expensive. The cost and size of the ground stations required for satellite communication is in inverse relationship to the power, and cost, of the satellite. Because satellites of the sixties and seventies were small and relatively weak in power, the earth stations required to transmit and receive signals to and from them were necessarily large and costly.

By the 1980s satellites and launchers had become big and powerful; the size and cost of the earth sections of the communication chain UPLINK-SATELLITE-DOWNLINK declined. For the first time the antenna and electronics necessary for direct-to-home reception of satellite television were available to individuals in Western Europe at reasonably affordable cost. Between 1982 and 1988 a small but growing population of European viewers received satellite television directly via personal TVROs (Television Receive Only Earth Station). In 1988 the first European Direct Broadcast Satellite, Astra, was launched, making possible direct reception of satellite television at an affordable price.

This European experience followed in the wake of developments in North America. In 1975 the US cable television programme company Home Box Office (HBO) began distributing its movie service to cable head ends (that is, to cable television operators for re-distribution to cable subscribers) via satellite. HBO 'wholesaled' its signal to cable 'retailers' via satellite in an unscrambled form. In the United States and Canada this led to individuals, bars, hotels and clubs acquiring TVROs in order to 'eavesdrop' on the HBO signal and access its movies without paying the subscription levied on authorized viewers. In the United States authorized viewers accessed the signal via cable, and paid a subscription charge in order to do so. In Canada HBO was not available via cable and the only means of reception, via TVRO, was illegal. Nonetheless many Canadians acquired a TVRO simply in order to watch HBO. In both North American countries HBO was a powerful stimulant to the TVRO market and an important factor in the development of direct-to-home satellite broadcasting. However no US broadcasting satellites to date transmit at power levels sufficient to make possible reception with TVROs of 60 cm diameter or less (60 cm has now become the size of TVRO designated, somewhat arbitrarily, as dividing DBS from other satellite television services). Strictly speaking, in terms of WARC (World Administrative Radio Conference) definitions, a DBS or television satellite is one which transmits at not less than 65 dBw to an antenna of 90 cm diameter or less.

Canada established a number of landmarks in satellite communication. Services began early, with the launch of the non-geostationary Alouette satellite in 1962, which made Canada the world's third satellite-owning country, but have never been extensive. The Alouette was followed by the world's first geostationary satellite, Canada's Anik A-1 in 1972. Anik has been used only by the CBC (Canadian Broadcasting Corporation, Canada's federally funded national broadcaster), to interlink its terrestrial transmitters, not to directly distribute television services to viewers. After Anik came Hermes in 1976, which established the first legal direct-to-home satellite service.

Despite this head start, Canada's satellites have characteristically operated with substantial unused capacity – neither telecommunications networks nor broadcasters have found it advantageous to replace established terrestrial systems with satellites. In consequence most direct-to-home reception of satellite television in Canada has been the reception (legal and illegal) of signals emanating from the USA.

Canada's experience of trans-border spillover of satellite television (designated as a 'problem' by the Canadian state, but percieved as a benefit by many individual Canadian viewers) is thought by some European states to pre-echo the loss of communication sovereignty they anticipate will follow the general establishment of satellite television in Europe. But it is important to recognize that Canadian viewers have accessed satellite television transmitting US programmes not only directly from the USA (transmitted from US satellites such as that distributing the HBO service) but also from Canadian satellites which redistribute US television (such as the CANCOM service which carries US signals in order to induce Canadian viewers to orientate their TVROs to the orbital position of a Canadian satellite and thus promote reception of the Canadian services transmitted from the same satellite). The pervasive presence and significance of American television in Canada (whether or not distributed via satellite) cannot be understood as a simple case of loss of communication sovereignty and involuntary reception of unwelcome exogenous television.

From Telstar to DBS: three generations of European television by satellite

Television via satellite is received for final consumption in the home through a variety of delivery systems. Each generation of television satellite has had a distinctive form of linkage to final consumers.

The first generation of television satellites, such as Telstar, transmitted weak signals which could be received only by very large and costly receiving dishes, such as those operated by British Telecom at Goonhilly and Madley. Final consumers received television signals at home via a complex distribution chain. Transatlantic television signals, such as those relayed by Telstar from the USA to the UK, might have originated in a New York television studio and have been sent via landline to an uplink dish antenna such as that at Andover, Maine. From Andover the television signal would have passed to a satellite whence it would have been downlinked to a similar large antenna in the receiving locality. From there, via landline or terrestrial microwave, it would travel to a UK television studio for landline relay to a terrestrial transmitter and thence through over-the-air terrestrial broadcast to the television aerial on the roof of the homes of individual television subscribers. Much 'satellite television' is still of this kind, and has the status of a 'producer' rather than a consumer service: the satellite acts like a link between a factory and a wholesaler in a retail distribution chain. For example, it links broadcasters to the Olympic Games, or redistributes television news footage between broadcasters. Television satellites used today in such chains are now considerably more powerful than was Telstar and therefore do not require such gigantic dishes as they used to.

During the 1980s a second generation of satellite television became available to final consumers, this time without the intervention of a terrestrial broadcaster. But the signals from these second generation, 'telecommunication' satellites (that is satellites established in orbit primarily to interconnect public telecommunication operators but used for 'satellite television' in Western Europe from 1982 onwards) were seldom received directly by final consumers using an individual dish. More often such satellites fed television signals to large TVROs for *re-distribution.* The most common system of redistribution was, and is, via cable. But other means such as terrestial rebroadcast, SMATV (Satellite Master Antenna Television, *ie* a TVRO shared by a number of final consumers, for example in a block of flats) or MVDS (Microwave Video Distribution System, sometimes known as MMDS (Multipoint Microwave Distribution System), a form of terrestial re-broadcasting in a restricted geographical area, requiring a line of sight between receiver and re-broadcast transmitter) are also used.

Third generation high powered direct broadcast satellites have reduced the size and cost of the TVRO required and have made *direct* reception of satellite television more attractive to final consumers than it was before. The most important European DBS (though not a 'true' DBS under WARC criteria) is the Luxembourg Astra satellite which transmits a bundle of services to television viewers in Western Europe. The main target audience for Astra services are Anglophone viewers in the UK and Ireland. To them seven channels (eight programme streams) of television are transmitted by Sky Television, W H Smith Television and MTV. There are also services directed to Scandinavian audiences transmitted from Astra. These DBS services are, like their forerunners, often re-distributed to final consumers via cable (and other media) but have also made it possible, for the first time in Western Europe, for a large audience to experience direct-to-home satellite television at affordable cost.

2 The shaping of satellite television

Satellite television is a complex phenomenon the evolution of which has been shaped by the interaction of a number of separate factors. I discuss below a number of the most important. As will be seen, understanding the dynamics of each of these distinct factors (such as the cost and reliability of satellite launch systems, and the cost/benefit trade-offs involved in different signal encoding/encryption systems) is not a straightforward matter. Still less is understanding the interrelationship between all these factors when they combine to produce particular outcomes. But to answer questions such as:

– Will satellite television succeed?
– What effect will satellite television have on viewers?
– What effect will it have on terrestrial television?

the following terms and ideas need to be understood, both as discrete and as interacting elements in the total phenomenon of satellite television.

Footprints

The 'footprint' of a satellite is the geographical area on the surface of the earth where signals transmitted from the satellite can be received. The transmitter on a satellite can be engineered to concentrate power on a desired geographical area, permitting reception with a small receiving antenna dish and relatively unsophisticated receivers. Or it can be engineered so as to spread the footprint more widely, thus necessitating bigger and more expensive antennae and receivers, thereby trading off signal strength against the extent of signal propagation. In theory, signals are capable of being received at any location with a line of sight between receiver and transmitting satellite, so long as the receiving antenna is sufficiently large and receiver circuits sufficiently sensitive. Reception in areas outside the designed footprint can be achieved, but only with commensurately more costly equipment. Clearly reception is cheaper and more reliable within the designed reception area established for a particular satellite than outside it.

The footprints for Western Europe's most important existing telecommunication satellites and for actual and planned DBS are shown below. The numbers on each 'contour' line of the footprint denote signal strength – the so called EIRP (Equivalent Isotropic Radiated Power). The higher the number, the stronger the signal and the smaller the receiving antenna required for satisfactory reception within the footprint. There is an inverse relationship between the strength of a signal and the area over which a footprint falls. Like light, signals from satellites can be 'focussed' in spot beams to concentrate power in designated service areas. Locations outside the footprint of such spot beams require larger antennae and more sensitive receivers than in locations within the footprint of the spot beam. Reception outside the designated area is, therefore, likely to be more costly than within it.

The Telecom, Intelsat and Eutelsat satellites are low-powered telecommunication satellites which carry television signals subject to pre-emption in favour of telecommunication services. Astra, launched in late 1988, is the first true television satellite to establish service to the UK. It is medium-powered (*ca* 47 watts transmitter power) and therefore not a true high-powered Direct Broadcast Satellite. Whereas the Marco Polo satellites from which BSB (British Satellite Broadcasting) began broadcasting to UK viewers in 1990 is a true DBS. The first BSB satellite was launched in August 1989 and has a designed transmitter power more than twice that of Astra.

Cable

Cable is the most important system for the re-distribution of TV signals from the satellite to final consumer. Most final consumers depend on it for access to low powered satellite television, and for many it will be the preferred medium of access to high powered satellite television. Yet it is expensive and unlikely to extend, at best, beyond the West European urban areas. UK estimates suggest that 50 per cent of the UK is unlikely ever

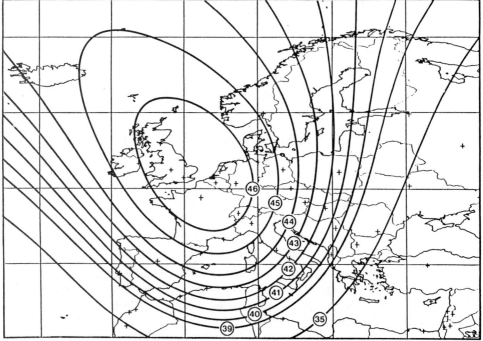

Eutelsat 1 F1, 13° East (spot West)

SATELLITE TELEVISION IN WESTERN EUROPE

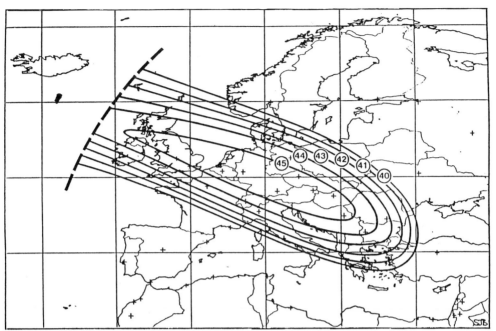

Intelsat VA 60°E, (EIRP contours spot West, half transponders)

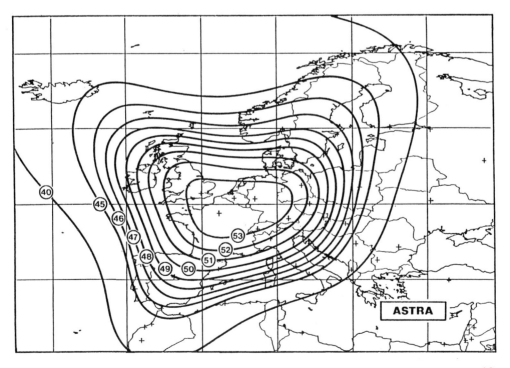

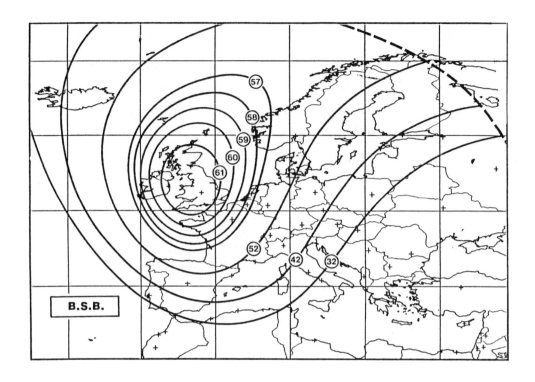

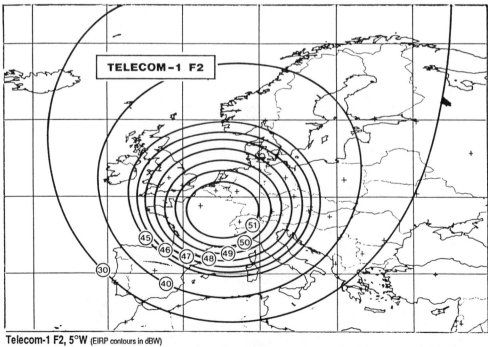

Telecom-1 F2, 5°W (EIRP contours in dBW)

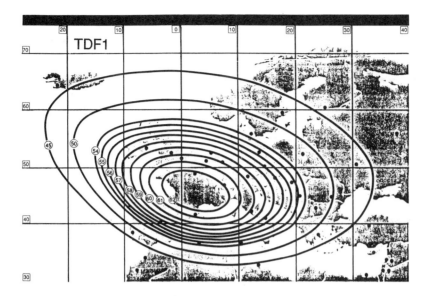

to be cabled and only very optimistic forecasts predict as much as 25 per cent cabling of West European households by the year 2000.

The two distribution technologies, satellite and cable, are customarily and misleadingly considered as rivals. Though rivals in some circumstances they are at present best considered as complementary and interdependent systems of television distribution.

For direct reception of signals from several satellites a viewer will require several dishes, or one expensive steerable dish, and an array of receivers capable of 'translating' transmissions, which may be encoded differently, to a single standard for display on a TV receiver. If signals are 'scrambled' to inhibit unauthorized reception, as subscription services characteristically are, a separate de-scrambler is also required for each encryption system used. UK viewers wishing to receive services from Astra and BSB therefore had to acquire and site two dishes (in addition to the aerials they already have for reception of terrestrial television), and attach to their televisions a separate receiver for each satellite, and a separate de-scrambler for each subscription service desired.

Reception of satellite and terrestrial television through a single cable may therefore prove to be an attractive alternative to direct reception for many viewers. Cable operators are able to establish the necessary 'dish farm', translate and de-scramble signals to a common standard and distribute all signals to subscribers through a single network.

Cable operators are also able to distribute non-broadcast television signals to subscribers. Signals may be originated at the cable 'head end' (*eg* from a community television studio), and the cable network may be interconnected to other networks enabling signals to be exchanged and widely distributed. Video tapes can be 'bicycled' between head ends (the UK cable film channels 'Bravo' and HVC – Home Video Channel – are operated in this way).

Cable distribution also offers satellite television programme providers the advantage of billing subscribers for subscription services and in feeding back information about consumption levels (though this benefit is most pronounced in advanced 'interactive' cable systems of which there are few extant in the UK).

However the full potentiality of cable networks is not always realized. There is little interconnection between cable networks, bicycling of tapes is inconvenient and expensive, and most importantly, cable networks are very costly and the asset created by investment in cable is illiquid. New cable networks are high risk investments and therefore few have been constructed except in circumstances where either public bodies have taken the investment risk or networks are used to redistribute terrestrial television from neighbour countries. Neither the UK policy of 'leave provision to the market' nor the state sponsored initiatives of West Germany and France have successfully established comprehensive cable networks, or attracted many subscribers to the networks that have been established. The success or failure of cable and satellite, whether in combination or as stand-alone services, depends on the cost and attractiveness of the services they offer to consumers. Too few consumers have so far judged returns to be sufficient to outweigh costs.

Cable television in Western Europe

The extent of cable penetration in Europe is difficult to estimate. However, two broad patterns of provision can be identified.

Firstly, the majority of networks are relatively 'old' narrow bandwidth cable systems developed for re-distribution of terrestial television. These are most evident in the small countries of Europe and are now also used for re-distribution of satellite television. However, they often lack the bandwidth to re-distribute all the satellite signals currently on offer (and for which there is a potential demand), still less the increasing range of new channels likely to be offered on new satellites.

The percentage of West European households subscribing to cable television networks in 1984

Country	Percentage
Belgium	81 per cent
Netherlands	74 per cent
Switzerland	47 per cent
Ireland	25 per cent
Norway	14 per cent
Denmark	10 per cent

In all other Western European countries cable penetration was below 10 per cent.

These smaller countries established narrowband services in order to relay to viewers terrestrial television signals from a neighbour country with a comprehensible language and a better funded television system. This rationale does not exist in the large West European states. Consequently states such as France, West Germany and the United Kingdom do not have extensive narrowband cable networks.

Instead, in the second pattern of provision, each of these larger states has seen recent initiatives to establish new, broadband, cable networks. Broadband cable (that is, cable systems with bandwidth sufficient for thirty or so television channels) has been promoted in these states for industrial policy reasons. There is a belief that broadband will stimulate national hardware businesses (in particular, manufacturers of fibre optic cable), and another belief that the desired economic transition to the status of an 'Information Society' is conditional on the existence of a national broadband telecommunications system. The assumption is that such a system can only be 'pulled through' by incorporating entertainment services in its offer. This is not the place for discussion of the broadband initiatives (see, *inter alia*, Collins 1986) but it is striking how unsuccessful they have been, whether market-led (as in the UK) or promoted by state or quasi-state agencies (as in West Germany and France).

In the late 1980s UK had about 273,000 homes which used cable as a means of accessing non-terrestrial television channels, though cable networks pass about 1.5 million homes (*ie* 1.5 million homes represents the potential subscriber market). This means that only about 18 per cent of potential cable subscribers actually subscribed to cable.

Most UK homes passed by cable are passed by old limited-capacity networks (built to relay terrestrial television in areas which had poor over-the-air reception) which, in some cases, have capacity for only one additional non-terrestrial service. However, the percentage of subscribers as a proportion of homes passed by cable does not differ greatly between narrowband and new broadband systems. It is low in both cases, less than half that required by operators for an economic return on investment.

(The statistics on availability and uptake of cable services are not particularly reliable; for obvious reasons they fluctuate from day to day. The difficulty of establishing authoritative statistics for cable television is common to the whole field of the new information and communication technologies, including satellite television. All statistics cited in this study should be used with caution.)

The Cable Authority is reported to estimate that 6.5 million homes will be passed by cable in 1992 and that 10 million homes will be passed in 1995 (*Financial Times*, 14.3.1989, p. 22). These estimates appear very optimistic when considered in relation to the number of subscriptions achieved in established systems (which are likely to have been built in the most attractive franchise areas). In January 1989 the UK had 11 broadband cable franchises in operation, 20 franchises had been awarded, six applications were under review by the Cable Authority, 10 areas had been advertised, and four decisions on licence were 'pending'. Of the 20 franchises awarded at least 10, awarded before January 1988, had undertaken no construction work (*Screen Digest*, Feb 1989, p. 40). The Cable Authority's estimates in 1989 were indeed optimistic. In October 1991 less than 1.2 million UK homes were passed by broadband cable (and less than 2.2 million by narrow, broad and SMATV [satellite master antena television] systems). Broadband penetration was 18.9 per cent of homes passed in 44 operating broadband franchises (*New Media Markets*, 21.11.1991, p. 5).

In contrast to the UK, French and West German cable services have been established by, or in conjunction with, public sector bodies. They have met with little more success than has been achieved in the UK. The French plan (Plan Cable) envisaged 1.5 million homes connected by fibre-optic cable by 1985 and an increment of 1 million homes annually thereafter. By the end of 1985, only 200,000 homes were connected. Penetration rates for cable in France averaged 19 per cent of homes passed, somewhat better than in the UK but still far from generating adequate returns for operators (*Mission Cable* report cited in *Cable & Satellite Europe*, Jan 1989, p. 15). Cable service in West Germany has been established more rapidly than in France or the UK and has achieved a higher penetration of homes passed; in 1987 34.4 per cent of West German homes were passed by cable of which 36.1 per cent subscribed to it (Deutsche BundesPost Press Release, 29.1.1988).

Terrestrial re-broadcasting of satellite television

Terrestrial re-broadcasting of satellite signals is an important and growing form of re-distribution. It does not necessitate use of capital-hungry cable networks and, as frequency planning authorities 'find' frequencies for additional terrestial services, and technical developments (*eg* MVDS) make usable frequencies hitherto useless, terrestrial re-distribution of satellite signals is likely to grow in importance.

MVDS (Multipoint Video Distribution System) or sometimes MMDS (Multipoint Microwave Distribution System) uses radio frequencies above 1 GHz to send broadcast signals from terrestrial transmitters to individual antennae linked to the television (and radio) receivers of final consumers. Like satellite television MVDS is a technology that uses radio frequencies which were formerly unusable for broadcasting. But unlike satellite television, which has the property of distributing signals across a very wide geographical area (all locations with a line of sight to the satellite), MVDS offers reception only in limited geographical areas.

The size of an MVDS reception area depends on numerous factors, of which the transmission frequency is particularly important. Signals at the high frequencies proposed for MVDS transmissions (because they occupy a relatively little used section of the radio spectrum and offer the possibility of carriage of wideband signals delivering superior picture and sound quality) attenuate rapidly, and travel in lines of sight.

MVDS is therefore a distribution medium that is likely to be difficult to establish satisfactorily in cities, where high buildings obstruct transmission paths, though advocates of MVDS express confidence in 'beam bending' technologies which are claimed to solve such problems.

Practical realization of an MVDS distribution system is likely to be in conjunction with cable rather than as a stand-alone system. Areas 'shadowed' from MVDS signals by high buildings and local topography are likely to be cabled. Indeed the cable industry in the UK is an important lobby for MVDS, which it advocates in order to bring 'cable' services to subscribers without the necessity of investment in trenching, cable laying and cable hardware.

There are no European MVDS services currently in operation, but the Republic of Ireland plans to establish an extensive MVDS service. The UK Government has published a consultant's report (DTI, 1988) which argued that MVDS services in the UK were technically feasible. The UK Broadcasting White Paper, 'Broadcasting in the '90s', states that the Government is proposing measures 'to facilitate the use of cable and MVDS for the local delivery of services' (Home Office, 1988, p. 27).

In this instance, as in others, we see that satellite television is not a stand-alone technology but has a complex technological, economic, and social interrelationship with other systems of signal distribution. If MVDS/MMDS services are widely established then low powered telecommunication satellites will be able to 'wholesale' signals to 'retail' redistribution transmitters which will then broadcast them for final consumption. Such services may well be cheaper and more satisfactory to final consumers than direct-to-home broadcasting by powerful DBS sustems. Touche Ross, (the consultants commissioned by the UK Department of Trade and Industry to report on MVDS) tabulate the relative costs of different television distribution systems as follows:

Costs per subscriber comparison

Take up	MVDS low frequency	MVDS medium frequency	Cable	DBS high power
15%	£232	£515	£1667	£330
25%	£227	£502	£1000	£300
50%	£224	£492	£500	£275

They comment:
> 'MVDS, at low frequency has the lowest per subscriber cost (assuming identical penetration rates) among the three forms of video distribution systems ... . In comparison to high power DBS, MVDS requires less up front cost, can be deployed faster, and could supply more channels to the home'. (DTI, 1988, p. 102)

MVDS is not the only form of terrestrial redistribution of satellite television signals. Indeed conventional over-the-air broadcasting from, and to, ground transmit and receive aerials is widely used as a means of disseminating satellite television. Many West German television viewers' access to the satellite television services RTL Plus and Sat Eins is via conventional terrestrial re-broadcasting. And in Greece the Greek public broadcaster, ERT, uses terrestrial re-broadcasting to transmit signals from low powered television satellites to Greek television viewers. Although there are no terrestrially re-broadcast satellite television channels in the UK, UK viewers do access satellite television via terrestrial re-broadcasting when an authorized UK broadcaster, such as the BBC or an ITV contractor, takes a satellite feed (*eg* from the Olympic Games) and re-broadcasts it to viewers.

The consequences of the differences between distribution systems

Providers of television programme services on low powered telecommunication satellites have tended to market their programmes to cabled areas – to the small countries of Western Europe, and to areas where terrestrial re-broadcasting has been available, such as West Germany. Providers of programme services transmitted from DBS satellites tend to direct their services to the large, uncabled, Western European countries such as the UK. The estimated penetration of satellite TV services prior to establishment of DBS is shown below. The small, extensively cabled countries have high access to satellite

Access to satellite TV channels by channel and country

	3Sat	Filmnet	RAI	RTL Plus	Sat 1	Sky	Super Channel	Teleclub	TV5	BBC	Kindernet	MTV
AUSTRIA	450,000				400,000	450,000	358,475	413,000		13,428		
BELGIUM		75,000	2,500,000		100,000*		1,367,259	1,759,000		1,539,666		443,609
DENMARK		2,000					496,984	483,000		197,674	140,000	323,997
FINLAND		9,000					379,376	395,000		181,050		191,684
FRANCE							94,395	49,000		99,995		35,180
GERMANY	3,800,000			3,800,000	3,608,000	3,977,040	3,968,000	1,000		3,198,000		595,000
IRELAND							304,390	296,000		27,000		259,000
LUXEMBOURG		50,000					86,518	91,000		65,654		
NETHERLANDS		79,000		100,000*			3,714,635	3,699,000		2,762,587	50,000	1,523,161
NORWAY		3,000					355,723	400,000		207,717	30,000	12,000	22,000
SPAIN							217,610	32,000		5,000		
SWEDEN		45,000					526,678	486,000		408,668	5,000	454,753
SWITZ	670,000		1,025,000		115,000	1,431,960	1,167,000	63,000	696,507			148,900
UK							300,483	152,000		7,821		106,013
TOTALS	4,920,000	213,000	3,575,000	4,400,000	4,173,000	13,874,853	13,390,000	64,000	9,410,767	175,000†	62,000	4,103,297

† Figures for BBC are estimates

Children's Channel	CNN	Arts	Premiere	Lifestyle	Screensport	BR3	Eins Plus	Eureka	Tele 5	WDR3	NRK	La Cinq	M6
		358,475											
		1,367,259											
		496,984		7,000	7,000								
	441	379,376			2,050								
	11,847	94,395			28,000							144,507	134,507
	275,178	3,997,040	365,000	365,000		3,609,000	3,500,000	3,276,000	3,460,000	3,609,000			
250,000	4,520	304,390	268,000	268,000									
		86,518	160,000	160,000									
650,000	597,800	3,714,635	34,000	34,000									
	895	355,723	14,350	14,350							420,000		
		217,610	130	130									
90,000*	24,873	526,678	16,500	22,000									
		1,431,960											
170,000	33,858	300,683	103,000	71,000	115,000								
1,160,000	949,312	13,611,526	103,000	835,980	1,015,530	3,609,000	3,500,000	3,276,000	3,460,000	3,609,000	420,000	144,507	134,507

* Figure represent Scandinavia — no breakdown available

Source: *Cable and Satellite Europe*, November 1988, pp. 70–71.

television via cable; West Germany has high access to satellite television via terrestrial re-broadcasting and via cable. Elsewhere there is low access to satellite television.

Satellite operators

The satellites from which television signals are delivered are operated by national agencies, for example France's TDF1 and Luxembourg's Astra, and also by international and regional satellite operators. Intelsat and InterSputnik are international agencies, while Eutelsat is based solely in the European region. They all operate satellites which deliver television services to Western Europe.

Intelsat (International Telecommunications Satellite Organization) was established in 1964 by 11 member states. Membership is open to all members of the ITU (International Telecommunications Union). Intelsat (which has its headquarters in Washington, DC) is predominantly a western agency: although the People's Republic of China and Yugoslavia are members, Cuba and the Warsaw Pact states are not. Intelsat's founding agreement states that its purpose is to provide: 'the space segment required for international public telecommunications services of high quality and reliability to be available on a non-discriminatory basis to all areas of the world'.

By 1987 Intelsat had 114 members, each member state nominating a telecommunications operator as its member representative. BTI (British Telecom International) is the member nominated by the UK. The ownership of Intelsat varies as the proportion of the shareholding, and thus ownership, of each member varies in proportion to members' usage of Intelsat services. However in the mid 1980s the USA (Comsat) owned about 25 per cent, the UK (British Telecom) about 14 per cent, and France, Japan, West Germany, Australia, Canada, Italy, Saudi Arabia and Spain between 5 per cent and 2 per cent. Intelsat deploys 15 satellites, 600 earth stations and 165,000 voice and data channels. This strength makes it the dominant agency providing television satellite services.

Intelsat's transatlantic satellite television service began in 1965. In 1986 Intelsat began to lease satellite transponders for domestic television services (*ie* services originating in and designed for consumption in a single state) as well as continuing to do so for its long established international services. In 1987 twenty six states rented transponders for domestic services (but not necessarily television services) from Intelsat.

1986-7 Intelsat revenues

73.1%	Telephony
8.1%	International data
5.1%	Domestic leasing
5.1%	Occasional TV
2.8%	Inmarsat (Maritime Navigation system)
2.5%	International TV
0.9%	Intelsat business
2.4%	Other

Intelsat currently raises only a relatively small proportion of its revenue from television satellites, but estimates that by the early 1990s this proportion will have risen to 20 per cent.

Intersputnik was established in 1971 by nine member states which feared dominance of the United States in Intelsat and grew to a membership of 14 states. There is some overlap of membership and facilities usage with Intelsat (Afghanistan and Vietnam are members of both organizations) and some states use both Intelsat and InterSputnik facilities but are not members of Intersputnik (Algeria, Angola, Ethiopia, Iraq, Libya, Nicaragua). But in spite of these crossovers, Intersputnik is best understood as the Socialist countries' equivalent to the 'capitalist' Intelsat.

Eutelsat (European Telecommunications Satellite Organization) was the first regional satellite system established anywhere in the world. It was founded in 1977 by CEPT (Conference of European Postal and Telecommunications Administrations) and has 26 member states in Western Europe including Yugoslavia, Finland and Turkey. It is fundamental to the operation of Eutelsat in that it deals only with the national telecommunication authority of its member states, not with governments. Its first satellite, the OTS (Orbital Test Satellite) was launched in 1978. Eutelsat now has four operating low powered satellites, of which three are used, at least in part, for television. Eutelsat plans to launch four powerful DBS satellites in the 1990s. 70 per cent of Eutelsat's income (1988) is reported to derive from satellite broadcasting revenues (*Communications Week International*, 20.3.1989 p. C5).

Société Européene des Satellites (SES) was franchised for 22 years by the Government of Luxembourg in 1985 to operate a satellite system. Its Astra satellite launched in November 1988 has 16 transponders. The shareholders in Astra are limited to a maximum holding of 10 per cent. The major shareholders include Luxembourg, Belgian and German banks and Thames Television and Kinnevik (the operators of ScanSat). Other UK television companies, Ulster Television and Television South West, have small stakes.

PanAmSat, originally authorized to provide service in North America, gained access to Europe in reciprocity for the UK telecommunications company Cable and Wireless' access to the United States. PanAmSat is managed, and was established, by Réné Anselmo, a close associate of the owners of the giant Mexican television corporation Televisa.

The entry of SES and PanAmSat into Europe has broken the control of national telecommunication authorities (PTTs) over access to satellites in Western Europe. Unlike Intelsat and Eutelsat, which are the children of the PTTs (and through them the national governments which characteristically control telecommunications in Europe), SES and PanAmSat are commercial organizations and have significantly weakened the ability of governments to control provision of satellite broadcasting and telecommunication services in Europe.

In 1986 British Satellite Broadcasting (BSB) was awarded an exclusive licence by the Independent Broadcasting Authority to operate a UK DBS for 15 years. BSB's satellite was launched in August 1989 with five transponders, of which three are definitely to be used for BSB's licensed services. The other two transponders are the subject of licence applications at the time of writing: BSB itself is an applicant.

West Germany and France have both launched communication satellites that are used, or planned, for broadcasting purposes. Each state has also planned DBS services. West Germany intended to do so from its TV Sat, but this failed to operate after its launch in 1988. France's TDF1 satellite, however, has now been established in orbit and commissioned.

Orbits and frequencies

Geostationary orbits and transmission frequencies for communication satellites are finite resources allocated by international agreement. In some localities, such as the USA, demand for orbits and frequencies exceeds supply. All slots in the geostationary orbits available to US operators of medium powered satellites are now filled. Communication satellites have made possible the use of frequencies in the Super High Frequency range that were hitherto effectively unusable for long distance broadcast, or point-to-point communication, thus easing pressure on other sections of the radio spectrum; but nonetheless the spectrum remains a finite and relatively scarce resource. Indeed satellites sharing adjacent orbital positions may potentially interfere with each other's signals if they transmit or receive on similar frequencies. This problem of potential interference is mitigated in practice by widely separating the orbital locations of satellites transmitting on adjacent frequencies, and by ensuring that signals are differently polarized, either horizontally or vertically. Individual satellites must also use different frequencies for their transmit and receive functions (otherwise uplinked signals would interfere with downlinked signals and vice versa). Thus a satellite may be described as being in the 14/12 GHz range; meaning that its send frequency is in the 12 GHz range and its receive frequency in the 14 GHz range.

The high frequencies used by communication satellites have particular characteristics. One is that they can carry large amounts of information, and effectively offer the only possibility of a broadcast medium for High Definition Television, HDTV, which requires considerably more spectrum bandwidth than do established television services). Another is that they travel in straight lines (requiring very accurate alignment of antennae to satellites if poor reception and interference is not to be suffered), and being easily obstructed by obstacles between sender and receiver, such as foliage and even rain and snow clouds.

The agency through which the resources of orbits and frequencies are allocated is the International Telecommunication Union (ITU). The ITU was established in Paris in 1865 (as the International Telegraph Union) and now has the status of a United Nations agency charged with regulation and development of telecommunications internationally. The ITU has been the locus for continuing international disagreement about the basis on which the finite resources of radio spectrum and satellite orbits should be allocated.

Conflict has recently focussed on how to balance the principles of equity, and ability to use resources. If, for example, an orbital position is allocated on a so called *a priori* basis (that is independently of need and ability to use the resource) to a state with neither the financial nor the industrial capacity to use the orbit, then the resource is likely to be wasted. If on the other hand orbits and radio frequencies are allocated to states on the basis of 'first come first served' then less-developed states are likely to find all orbits and frequencies occupied when they acquire the capacity to use their 'share' of the frequencies and orbits which such states generally deem to be a global resource.

Further conflicts have arisen, including one over the claim of equatorial states that their sovereignty extends upwards from the earth's surface to infinity. Therefore, equatorial states argue, geostationary satellite orbits are national resources (like airspace) and should not be allocated as a global resource. The equatorial states formulated their claim to sovereignty over the geostationary orbit in Colombia in 1976 and issued the Bogota Declaration to assert their claim.

Until the early 1970s the ITU allocated resources on a 'first come first served' basis. In 1973 it formally recognized the right of all states access to the geostationary orbit, but effective implementation of this policy has remained unrealized. In 1977 regulations for broadcasting satellites were established at a WARC (World Administrative Radio Conference) convened for that purpose. Frequencies and orbital slots for European broadcasting satellites, including DBS, were established there. The 1977 WARC was followed in 1979 by another, bigger WARC on general spectrum allocation, which in turn was followed by an ITU orbit conference (convened in 1985) established to find: 'an acceptable way of guaranteeing in practice to all countries equitable access to the geostationary satellite orbit and the frequency bands allocated to the space radio communication services, whilst reconciling that objective with the efficient and economic use of these natural resources' (Howkins and Pelton, 1987, p. 35)

The conflicts of interest negotiated in the fora of the ITU (and also contested in the United Nations and UNESCO) have never been permanently resolved. New technologies, new needs and new configurations of interest among the 162 member states of the ITU make any allocations of the ITU subject to revision and renegotiation. In particular the ITU (and other UN agencies and institutions) has become the site of political conflict between states with different and mutually antagonistic interests over the natural resources of the geostationary orbit and the radio frequency spectrum. Rather than being, as it once was, a regulatory and organizational body acting to administer a consensus established among its members, the ITU has become a forum for conflict as consensus about the nature of equitable and efficient allocation of resources has disappeared. But in spite of its 'politicization' the ITU has been remarkably successful in creating, and maintaining, an international regime that recognizes the mutuality of the interests of member states. That is, in having a spectrum and orbit regime that realizes the members' *common* interests such as non-pollution of the spectrum and maximization of spectrum utilization, rather than the individual interests of particular member states.

Even so, the introduction of broadcasting satellites has intensified concern over control of orbits and spectrum, and over national sovereignty. For broadcasting satellites propagate their signals across national frontiers and, whether for the first time or in a more intense form than heretofore, expose television viewers to exogenous signals. These signals, it is widely believed, have an unprecedented social and cultural impact on receiving populations, and therefore require regulation not hitherto found necessary for earlier forms of crossborder signal spillover. The ITU is likely to become a forum for increasingly acrimonious conflicts between members as the economic importance of orbits and frequencies increases, and as concern grows over transborder spillover and 'loss of communication sovereignty'.

Satellite orbits and Western Europe

Orbital locations for the principal West European fixed service satellites used for television are:

Orbital position (in degrees)	Satellite
63 °East	Intelsat VAF12 (principally German language services)
13 °East	Eutelsat IF1 (principally German language services)
50 °West	Telecom IF2 (principally French language services)
27.5 °West	Intelsat VAF11 (principally English language services)

DBS orbits have been allocated as follows:

Orbital position (in degrees)	Country
5 °East	Cyprus, Denmark, Finland, Greece, Iceland, Norway, Sweden, Turkey
1 °West	Bulgaria, Czechoslovakia, East Germany, Poland, Romania
7 °West	Albania, Yugoslavia
19 °West	Austria, Belgium, France, West Germany, Italy, Luxembourg, Netherlands, Switzerland
23 °West	USSR
31 °West	Ireland, Portugal, Spain, UK
37 °West	Andorra, Liechtenstein, Monaco, San Marino, Vatican
44 °West	USSR

The closer satellite orbital positions and transmission and reception frequencies are to each other, the more sensitive the receiver required for satisfactory signal reception. In turn the size of antenna and sensitivity of receiver are inversely related to power of transmission. There is some uncertainty about the size, and hence cost, of antennae required for different satellite services. But for medium powered DBS services such as Astra (transmitting with a power of 47 watts) viewers within the core area of the satellite's footprint (UK, France, West Germany, Benelux) will, it is claimed, receive a good quality signal with a 60 cm diameter dish. Outside the core reception area of the footprint a larger dish (or a more sensitive receiver) will be required. Astra viewers in the northern latitudes of the UK are recommended to use a 80cm dish. For services transmitted from higher powered satellites, such as BSB or TDF1, smaller dishes are likely to suffice.

Transponder capacity

The term 'transponder' is loosely used to signify the channel, and hardware on the satellite, used to receive an uplinked signal from earth and to downlink the signal back to earth. The bandwidth required for one transponder is *ca* 40 MHz.

It is difficult to predict future supply of transponders in Europe. Supply is conditional on successful launches and working satellites. Experience (of Ariane and Shuttle launcher failures and the write-off of the West German DBS, TV Sat) demonstrates how uncertain these factors are.

Demand is similarly hard to predict, since alternative delivery systems for both telecommunication and television services are available: co-axial and fibre optic cable, terrestrial broadcasting and microwave. Ivor Cohen (formerly managing director of Mullards) has stated: 'cable and satellite entrepreneurs face enormous financial risks because of competition from less costly technologies such as conventional terrestial television' (*Financial Times*, 19.2.1988, p. 8).

However it is clear that there is currently an excess of supply of transponders over demand for them and that possible growth in demand (*eg* from UK ITV companies such as LWT and Thames Television that promised to move their transmissions to satellites if they lose their ITV franchises in future franchise auctions. Thames did lose its franchise and has reaffirmed its intention to establish a satellite service) will be exceeded by growth in supply as more satellites (particularly DBS) are launched. This prediction is supported by a European Space Agency (ESA) estimate which suggests that: 'supply looks likely to grow faster than the fragile demand for it from the TV world. The result could easily be a glut of capacity that will concentrate the mind of those organizations involved in the provision of this satellite capacity' (ESA, 1986, p. 3).

However as higher powered DBS satellites become available some television services at present transmitted from low powered telecommunication (or FSS – fixed service satellites) satellites are likely to migrate to these new satellites in an attempt to reach a potentially large TVRO audience. The same ESA report (p. 7), estimates at 70 per cent the number of European homes in rural areas unlikely to be cabled. Such relocation of services is likely to increase excess capacity in the telecommunications low powered satellite sector (though transmission via Astra by Sky Channel, MTV and W H Smith's Lifestyle and Screensport channels has not yet led to cessation of long established transmissions from the low powered satellites Intelsat VAF11 and Eutelsat 1F4). By 1990, the ESA estimates, there will be over 100 transponders on low powered satellites (and perhaps 150 transponders in all).

Overcapacity on low powered satellites designated for telecommunications has historically been concealed by the satellites' utilization for television transmissions: the 'second generation' of satellite television in Europe. But television broadcasting by satellite will tend in future to migrate to high powered satellites specifically designated for TV use. The resulting surplus telecommunication capacity may be attractive to business users,

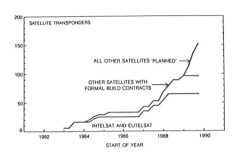 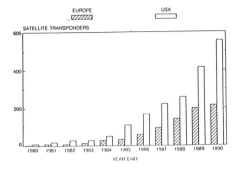

Source: Chaplin, 1986, p.23.

(who would otherwise use ISDN terrestrial facilities) if they are permitted to up-and-down link directly to and from the satellite, thus bypassing the costly 'last mile' from PTT earth stations to customer premises. Such developments, if permitted by regulators, will have profound effects on the tarriffing of telecommunication services but, unless extensive home working or greatly expanded business data transfer needs are established, satellite transponder capacity for telecommunications in Western Europe is likely to continue to exceed demand.

Transponder capacity for television is also likely to be oversupplied. The French DBS, TDF1, is perhaps the best example of a satellite undersupplied with programme content. It has only one non-terrestrial programme stream, La Sept, identified for it to carry. The European satellite programme is much more strongly 'pushed' by the lobbying power of the electronics and aerospace industry than it is 'pulled' by demand from television or telecommunication users.

1987 rates of *ca* $1.5 million pa for pre-emptible transponders on telecommunication satellites are likely to soften. The hopes of satellite operators that a substantial premium over existing rates could be established for higher power transmissions seem unlikely to stick. British Telecom International's (BTI) initial asking price of £4.95 million pa for transponders on the medium powered Astra satellite is reported (*Cable & Satellite Europe*, 7.1988, pp. 17–18) to have been negotiated down by News Corporation to £2.1 million per transponder per year for a non pre-emptible ten year contract.

CNN is reported to have paid $25 million for seven years' rental of a transponder on the low powered Intelsat VAF11 satellite (*Cable & Satellite Europe*, 7.1987, pp. 15–18).

Reception equipment

TVROs offer a means of accessing satellite TV to those who are unable, or unwilling, to subscribe to cable. Reception of transmissions from second generation telecommunication satellites via individual TVROs was relatively uncommon (a large and expensive dish was required) and cable was then the preferred means of accessing satellite television signals. The increased power of third-generation DBS satellites such as Astra and TDF1 have reduced the size and cost of TVRO required for satisfactory reception, and are likely to render TVROs the preferred means of accessing satellite television, provided that signals from only one satellite – or cluster of satellites in a single orbital position – are desired.

To receive signals from *one* satellite a fixed TVRO aligned to the orbital position for the preferred satellite suffices. But to receive signals from more than one satellite either a separate dish for each satellite, or a steerable dish, is required. And if signals from a non-geostationary satellite are desired a steerable dish is essential. To receive encoded or scrambled signals, decoders and descramblers are required. A UK viewer wishing to receive, for example, the US armed forces television (the 'best' from the US networks), CNN and Sky Channel would require either three dishes (for Intelsat VAF12, Intelsat VAF11 and Eutelsat IF1) or a steerable dish scanning between 60 °E and 27.5 °W. Also three systems, B-MAC, PAL and Orion, for decoding and descrambling signals. In order to view Sky Channel and the W H Smith channels (Lifestyle and Screensport) transmitted from the Astra satellite only one (relatively small) dish is required. However, when the channels are encrypted, as it is proposed they will be (Sky Movies and the W H Smith channels are scheduled for encryption in late 1989) then two separate decoders will be required. If W H Smith fulfils its intention to transmit using the D-MAC encoding system (rather than the PAL system used by the Sky channels and UK terrestrial television) then viewers will require an additional receiver for W H Smith channels. Should a viewer wish to receive English language signals from the different Astra and BSB DBS satellites, then two antennae will be required, and three separate decoders.

If consumption of a plurality of signals from a number of satellites is desired, then cable (if available) is likely to be a more attractive solution than numerous (or expensive steerable) TVROs and descramblers. However the capacity of cable systems is limited; even with the much hyped fibre-optic network the limits of capacity may quickly be reached and therefore not all desired channels may be carried on any single cable system.

Estimates of TVRO penetration vary greatly, whether current or future levels are at issue. The UK consultancy Logica estimates that the UK TVRO population in 1994 (that is five years after the planned establishment of BSB and Astra) will be less than 600,000 (*ie* less than 5 per cent of the potential) and that for Western Europe the figure will be 3.2 million. CIT (another UK consultancy) is even less optimistic, predicting a total European poulation of 1 million dishes by 1996. Ferguson, the UK based TV receiver manufacturer (owned by the French company Thompson) estimate a UK dish population of 1.5 million in 1990, a further 2 million in 1991 and a further 1.5 million in 1992.

Satellite TV contractors are more sanguine: BSB estimate 4 million TVROs in the UK by 1994 (*Cable & Satellite Europe*, 9.1987, p. 67), while Sky Channel estimates (*NMM*, vol. 7, no. 1, Jan 1989, p. 1) that the UK will have 600,000 TVRO homes, and Ireland 80,000, by the end of 1989. That is, the dish population Logica estimates will require five years to achieve, Sky Channel predicts will be achieved in one year.

Predicting the future is a hazardous activity and, depending on the initial assumptions that inform predictions (and perhaps the desires of the predictors), a plausible case can be made for almost any number of future viewers of satellite television. But although evidence is imperfect and contradictory, the publicly expressed expectations of the satellite television operators in the UK market, Sky Channel and BSB, seem unreasonably optimistic viewed in the light of the past consumption of European satellite television.

A telephone enquiry to the DTI (Department of Trade & Industry) in March 1988 revealed that fewer than 5000 TVRO licences had then been granted in the UK. At the time of the enquiry TVRO owners were required to purchase a dish licence but UK TVRO owners no longer face this requirement, so a potentially reliable index of dish penetration is no longer available to researchers.

Dish populations vary significantly between countries: Sweden, for example, is reputed to have 'one of the highest levels of TVRO installations in Europe' (Home Office, 1987, p. 27) but such comparisons may not mean very much. Overall, it is striking how few TVROs existed in Europe in the late 1980s, penetration of TVROs in Western Europe in mid-1988 was estimated to be:

Austria	6000
Belgium	Illegal
Denmark	8000
Finland	3600
France	10,000
Greece	1500
Ireland	2000
Italy	2000
Luxembourg	80
Netherlands	2000
Norway	3700
Portugal	2000
Spain	19,000
Sweden	17,500
Switzerland	1000
West Germany	19,500
UK	18,000

Cable & Satellite Europe, 10.1988, pp. 26–8.

Launchers

In order to position a geostationary satellite at 22,300 miles above the equator a powerful launcher rocket is required. Though the size and weight of satellite components are continually decreasing there is a constant countervailing demand for more and more powerful satellites. Powerful 'DBS' satellites are, all other things being equal, heavier than 'telecommunication' satellites and require correspondingly more powerful rockets to place them in orbit. Conventional rockets are costly and are used only once.

The alternative launch technology, the United States' space shuttle was developed in order to lower launch costs by recycling major parts of the satellite launch system. The US shuttles offer a payload weight of up to 30 tonnes. Shuttles have not directly established satellites in geostationary orbit but they are able to directly establish satellites into lower orbits. In this way they have offered significant economies in launching geostationary satellites. They take the satellite payload out of the powerful influence of the earth's gravitational field, thus permitting the use of a relatively lowpowered and cheap rocket to establish the satellite in its final orbit. This makes the operation cheaper than using expendable rockets.

The loss of the space shuttle 'Challenger' in January 1986 arrested use of the shuttle launch system, a system that had promised to lower the cost of establishing a television satellite in orbit. Shuttles were used for only 24 flights prior to the Challenger loss (with development costs, to 1985, in excess of $250 billion). The President of the USA stated that henceforth the shuttle would be used only for military payloads. Shuttle flights recommenced in early 1989.

Conventional, non-reusable rockets are, pending possible re-establishment of shuttle launches, the sole available means of launching television satellites. They are, however, unreliable and costly and do not have the high payload capacity of the shuttle. The United States, Soviet Union, China, Japan and, nominally, Western Europe (but, effectively, only France) have launchers capable of placing satellites in geostationary orbits. For West European clients some states, such as the Soviet Union, which have a demonstrated launch capability, are not realistic options as launch contractors. But it seems increasingly likely that China will become an acceptable launch contractor for Western satellites.

Launch capabilities

China's Long March rocket has an 85 per cent reliability record and the capacity to launch a 1400 kg load. The Soviet Union's Proton has a 92 per cent reliability record and a 2200 kg capacity. Japan is unable to offer launch capacity to non-Japanese customers before 1992 (due to technology licensing conditions from the USA) when a Japanese rocket H2 is scheduled for operation. The present Japanese H1 rocket has an excellent record of reliability and the capacity to place 550 kg satellites in geostationary orbit. H2 is planned to have a 2 tonne capacity.

The choices for European satellite launches have been between US and ESA (European Space Agency) launchers. The United States has three rocket launchers.

(a) The Titan III with 96 per cent reliability and a 5670 kg capacity.

(b) The Delta II with 98 per cent reliability and a 1819 kg capacity (this rocket launched the Marco Polo satellite used by BSB).

(c) The Atlas/Centaur with 95 per cent reliability and a 2358 kg capacity.

The European Space Agency's Ariane family (Ariane 1-3) has 79 per cent reliability over 19 launches. Payload capacity has grown with successive versions of the Ariane system. Ariane I had a capacity of 1850 kg, Ariane 2 a capacity of 2175 kg, and Ariane 3 a 2000 kg capacity. Ariane 4 is planned to have a launch capacity of 4200 kg.

The European Space Agency with Ariane has a competitive advantage over other launch systems in that its launch site (Kourou in French Guiana, an overseas Department of France) is closer to the equator than are the launch sites used by other launch contractors.

This means that more of Ariane's payload is usable for satellites and that, therefore, heavier payloads can be delivered by Ariane than by rockets of comparable power launched from other sites.

The Chinese and Soviet launchers have not been used by Western companies because of security considerations, though the USSR's Proton is potentially very attractive. It is reliable, very powerful (it can launch a satellite directly into geo-stationary orbit and avoid the need for a potentially troublesome apogee kick motor which can account for half the take off weight of a satellite) and has a frequent launch schedule. China now offers a service whereby launch clients retain control of their satellite at all times and are, therefore, less vulnerable to possible industrial or military espionage by the launch contractor. However no Western clients have yet used Chinese launchers.

Effectively, therefore, the choice for European services has been between the ESA Ariane and US launchers (of which Delta is the most reliable and Titan the most powerful) and are made on price/performance and political criteria. Ariane's historically inferior reliability in comparison to rival systems is clearly important, though its inferior performance may be due to fewer launches than its US competitors (the early section of a learning curve is always likely to be the most troublesome). Ariane claims that its mistakes are behind it, but some European customers have booked launch slots using rival US rockets (*eg* BSB on Delta and Eutelsat on Atlas Centaur) rather than risking failure with Ariane.

The failure of the US space shuttle removed an important supplier of launches from the market (at least until the backlog of US military launches has been reduced) and in consequence there is now a glut of customers for launches. It seems likely that all available launchers, including Ariane, will be booked until the 1990s and it is possible that Western clients crowded out from access to Western launchers will adopt a Chinese or, less likely, Soviet launcher (Sources: *Cable & Satellite Europe*, 10.1987, pp. 14–24; *Financial Times*, 27.1.1987, p. 20).

Insurance

An index of the uncertainty and hazards attached to the launch and establishment in orbit of a satellite is the cost of satellite insurance. By May 1986 cumulative losses suffered by insurers were estimated to be *ca* $923,500,000. Further launch failures and losses of satellites in orbit occasioned a rise in insurance costs (from *ca* 9 per cent in 1982 to 29 per cent in 1986) and a refusal by underwriters to insure more than approximately $100 million per launch. This meant that launches involving more than one satellite (as many do) might be uninsured for 80 per cent of potential losses.

The failure of Ariane 15 (carrying the ECS3 and Spacenet F3 satellites) cost the insurance market $150 million and engendered a further reluctance to extend cover to satellite launches. In consequence the agency operating the main expendable launch vehicle Ariane (and also the Chinese Long March rocket) now offers launches with insurance (*Cable & Satellite Europe*, 5.1987, pp. 40–45) by the launch agency. In spite of the poor record of successful launches in 1987 rates eased slightly to a premium of 20–25 per cent of the sum insured (*Cable & Satellite Europe*, 9.1988, pp. 56–61).

Transmission standards

'The European Commission has a goal of a unified European market in goods and services by 1992. Broadcasting is a component part of that unified market for services'. (Lord Cockfield, European Commission Vice President in *Europe*, Dec 1986, p. 49)

To achieve the unified European television market desired by the European Commission requires the harmonization of transmission standards for television in Western Europe. Two different line and coding standards for terrestrial television broadcasting currently prevail: SECAM and PAL. The existence of two incompatible standards creates barriers to a unified European terrestrial television market. (There are also differences within PAL: for example the UK and West German PAL systems do not relate sound and image signals identically.)

The inception of television in Western Europe after the Second World War saw adoption of a variety of line and coding standards. The UK used a 405 line system, France 819 lines and West Germany 625 lines. The introduction of colour television saw the UK adopt the PAL 625 line picture standard similar, but not identical to that used in West Germany. But France, though adopting a 625 line standard, used its own SECAM colour encoding standard. The different standards have prevented a common European hardware market developing in which, both for studio and domestic reception equipment, manufacturers could benefit from economies of scale.

However this difference in standards has not seriously inhibited trade in television programmes between European countries; for programmes have customarily traded in the form of 16 mm cinema films, and the falling cost of standards conversion has further reduced potentially important barriers to production and trade in video recordings. Where consumers have wished to receive broadcast programmes in more than one standard (*eg* in Francophone regions of Belgium) television receivers with multi-standard capabilities have been retailed. Though such multi-standard receivers are more costly than single standard sets, differences in standards have not prevented development of a European market in television software or hardware though they have raised the costs of the commodities internationally traded in that market.

Neither of the two established terrestrial broadcast standards – PAL and SECAM – is well adapted to satellite broadcasting. However, in default of anything better, the PAL standard was used initially. Accordingly a new European standard for satellite television, MAC (Multiplex Analog Components), has been developed (originally by the Engineering Division of the UK Independent Broadcasting Authority) for European Satellite Services. MAC has been designed as an 'evolutionary' standard permitting continued use of existing television receivers (with an additional decoder) but also initiating an evolutionary movement towards a desired European standard for High Definition Television (HDTV). But reception of MAC signals, though possible using existing receivers constructed for conventional terrestrial transmissions, requires additional expenditure by viewers on a MAC decoder.

Moreover MAC is ill-suited to many existing cable networks which lack sufficient bandwidth for satellite signals encoded in MAC. It is possible to distribute on existing cable networks the picture component of a MAC signal (in a degenerated but acceptable form) but not its digital sound. There is, therefore, a trade off between optimal sound and vision quality in signals encoded using MAC standards, and wide distribution of MAC signals, if existing distribution and reception equipment is to be employed.

However in spite of the adoption of MAC in the name of an advance towards an integrated European television market there is no unified MAC standard; rather a plurality of MAC standards have been developed which offer different balances of advantages and disadvantages.

C-MAC is the optimal MAC standard and was backed by the UK and Scandinavia for DBS services. Swedish and Norwegian public television use C-MAC for their satellite transmissions. It offers four-channel stereo sound or eight mono sound channels. C-MAC's plurality of sound channels offers the possibility of simultaneous transmission of a single vision signal and a number of sound signals. A TV programme could thus be transmitted in C-MAC and be simultaneously received in Germany, France, the Netherlands and the UK with all viewers able to hear the sound track in their native language.

B-MAC (used by Aussat in Australia and in Europe by the US armed forces television channel, by ScanSat for its low powered 'second generation' satellite service and by the UK satellite-delivered racing service to bookmakers) offers only three stereo sound channels.

D-MAC is a re-engineered C-MAC requiring less bandwidth than C-MAC, but still too much for most existing European cable systems.

D2-MAC halves the information carried by D-MAC resulting in a somewhat degraded picture and only two stereo (or four mono) sound channels in order to achieve compatibility with existing cable systems. There are two sub-categories of D2-MAC, one for terrestrial and cable distribution, and one for satellite distribution. D2-MAC has been adopted by the governments of France and West Germany for their DBS services and by ScanSat for its Astra transmissions. It is likely to become the dominant standard for continental european satellite television but probably not for services directed to the UK and Ireland where D-MAC is likely to prevail.

The MAC system, in spite of being developed to establish a unified European standard, promises to replicate the SECAM/PAL standards division. D2-MAC is clearly the standard best adapted to distribution of satellite signals over existing cable networks and to the early establishment of as large a market for satellite TV in Europe as is possible. Its lack of capacity for a number of sound channels recommends it to states that fear their national programme production industries and domestic broadcasters would be disadvantaged in competition in an integrated and barrier-free European broadcasting market. States that neither have an extensive cable infrastructure nor fear the economic and cultural consequences of market integration tend to favour C-MAC or D-MAC.

For DBS, C-MAC offers a standard that will optimize quality and reception in a plurality of languages via individual TVROs or advanced wideband cable systems. It is a better solution for direct-to-home broadcasting (and for evolution towards HDTV) than either established terrestrial standards such as PAL or any of the other MAC standards. But neither C-MAC nor D-MAC is likely to deliver early maximization of access to satellite television. To do that requires a signal capable of being distributed through established cable networks, which have a limited bandwidth capacity. Hence the adoption of D2-MAC and PAL standards by satellite television broadcasters. All MAC standards can readily be encrypted and decrypted without loss of quality.

Choice between MAC standards turns on the questions of wide or narrow bandwidth distribution, of advance towards HDTV, and of the desirability of an integrated broadcasting market. If a policy goal is utilization of existing cable systems of relatively narrow bandwidth then D2-MAC is to be preferred. If a policy goal is advance towards an optimal signal quality, and bandwidth constraints are unimportant, then C-MAC is to be preferred. If Europe's goal is a digital system with optimal quality (and therefore ability to compete with the Japanese MUSE HDTV system as a world standard), albeit requiring high bandwidth, C-MAC is to be preferred. In the long run C-MAC's sound channel capacity, greater than that possessed by other MAC standards, promises to better deliver a unified broadcasting market (as more sound channels in different languages can be offered simultaneously). However, not all European states and broadcasters welcome the idea of a unified and barrier-free broadcasting market and accordingly do not all favour C-MAC (or D-MAC). D2-MAC offers earlier realization of a relatively large market for satellite television than does C-MAC through redistribution of television signals via existing cable systems. It is also capable of future use as a digital standard and would promise slow but steady development of an integrated European television market.

The choice is therefore between an 'optimal' standard which will render many existing cable systems redundant, and will deliver an expanded market for satellite television relatively late, and a system well adapted to existing narrowband distribution but which promises to offer neither quality comparable to the Japanese HDTV system MUSE nor the plurality of sound channels likely to promote an integrated European television audience.

Effectively the choice is between D-MAC, a re-engineered C-MAC which consumes less bandwidth, and D2-MAC which best optimizes use of existing cable systems but does not offer the picture quality and plurality of sound channels available via D-MAC.

Once a standard has been chosen, implementation of the choice may prove difficult: though D2-MAC was chosen for French and West German DBS services, D2-MAC reception chips have proven very difficult to manufacture successfully.

Satellite television programmers have adopted different strategies with respect to MAC. The Sky Channel programme streams transmitted from Astra have eschewed MAC and are encoded and transmitted using PAL. Reception of Sky signals is therefore possible using the existing television receiver population without additional expenditure on a MAC-to-PAL converter. Sky signals also pose no technical difficulties for cable distribution.

BSB, on the other hand transmitted its services in MAC. It was required to do so by the European Community Directive of 1986 on satellite television broadcasting standards. The television satellites to which the Community regulations applied were those which qualified as television satellites under the WARC definition of a television satellite. Astra, from the point of view of WARC, is a telecommunications satellite. MAC offered BSB some benefits, better sound and image quality, more secure encryption of signals than is possible with PAL standards (and thus less signal piracy), the ability to transmit a plurality of soundtracks and to evolve towards an HDTV service, but the costs were high. BSB receivers were more costly than Sky receivers and thus Sky gained a crucial competitive advantage due to its choice of the PAL, rather than MAC, transmission standard.

Signal encryption

An issue related to transmission standards is scrambling, or encryption, of transmissions to inhibit unauthorized and/or unpaid reception of signals. There are a variety of scrambling systems in use, each requiring a different de-scrambler. For example, the BBC1/2 feed on Intelsat VAF11 27.5 °W is scrambled using a different system to that of Satellite Information Services carried on the same satellite. Sky Channel and FilmNet, both on Eutelsat 1. F1 13.0 °E, scramble signals using different systems. Final consumers wishing to receive both Sky and FilmNet require to purchase or rent a separate decoder for each channel. However all encryption systems used to date have been successfully 'hacked' and it is claimed that 'pirate de-scramblers are openly available' (*Cable & Satellite Europe*, 3.1989, p. 96). Piracy of the Premiere movie channel (admittedly using a simple encryption system) is claimed to account for between 40 per cent and 70 per cent of Premiere viewers.

The proposed English language subscription channels on Astra will use different encryption systems and protocols. A viewer wishing to receive both Sky Movies and W H Smith's Screensport (when encrypted) will require to buy two descramblers. The costs of decoder and de-scrambler rental (or purchase) and the inconvenience of a proliferation of separate boxes at the point of consumption will clearly inhibit viewers from consuming a plurality of different scrambled signals. A widely canvassed advantage of the MAC encoding system is its claimed superior scrambling system. However the study of signal piracy in *Cable and Satellite Europe* cited above (*Cable & Satellite Europe*, 3.1989, p. 96) expresses scepticism about the effectiveness of BSB's encryption system.

3 The funding of satellite television

Satellite television service providers can draw on three types of funding for programme services: state budgets (or licence fees), advertising finance and revenue from viewers' subscriptions. Of these, the first is straightforward and depends simply on political decisions taken by government agencies. Public funding (supplemented by advertising) sustains established satellite channels such as West Germany's Eins Plus and supported the aborted Eurikon/Europa experimental services run by consortia of West European public broadcasters between 1982 and 1986. However public funding is currently not very important for satellite television. Little new programming is transmitted on existing publicly funded channels (the West German channels – the most important channels financed from taxation – essentially relay programming developed and transmitted by public terrestrial broadcasters). The most important public sector initiative yet to be established, Europa/Eurikon, failed because public funding was not sustained. It is therefore not remarkable that no new public satellite television initiatives are currently being canvassed.

Accordingly only advertising and subscription finance, the funding mechanisms for most satellite televison channels in Western Europe, are extensively discussed in this study.

Advertising finance

Successful funding of satellite television by advertising depends on three factors.

> (1) The competitiveness of alternative advertising media in particular markets.

> (2) The existence of brands that are marketed within the reception area of the satellite service.

> (3) The relative attractiveness of the programming offered by satellite television in comparison to rival programme streams.

The leading UK advertising agency Saatchi & Saatchi Compton (1987, p. 1) has estimated that by 1995: '... new powerful satellites will have become established with, we estimate, around 35 per cent penetration of UK television households'.

Saatchi and Saatchi's assumptions were based on straight line projections of Cable Authority estimates of cable penetration and the assumption that penetration of TVROs in the UK will replicate that of VCRs. These seem very optimistic assumptions, as does Saatchi & Saatchi Compton's view that: '... satellite broadcasting across national frontiers – Pan European services – will predominate and be a key dynamic in our business' (Saatchi & Saatchi Compton, 1987, p. 1)

The West European television market has few brands common to all national markets within the product ranges customarily advertised on television. Therefore an advertising medium that delivers a transnational audience is of interest to very few advertisers. Moreover, a product that is available in more than one market may be used for different purposes in different markets and thus require different advertising campaigns for each distinct market. For example, cough sweets are sold in the UK to relieve winter ailments, but are sold in the Meditteranean to relieve sore throats caused by summer dust and aridity. The advertising campaigns required to promote cough sweets successfully in northern and southern Europe will therefore be different.

In contrast to Saatchi and Saatchi's bullish anticipation of an integrated advertising market Hans Thoma, the managing director of the German language satellite television channel RTL Plus, states: 'I have never believed in Pan European advertising' (*Cable & Satellite Europe*, 12.1987, p. 33).

The transnational European satellite television advertising market is estimated by an industry source (interviewed by the author in May 1988) to amount to about £20 million annually. It has grown from nothing in 1983. UK based channels, such as Sky Channel and Super Channel, which have been wholly financed by advertising, estimated that advertisements carried on their transnational services originated in roughly equal proportions ($1/3 : 1/3 : 1/3$) from US, Japanese and European companies. Sky Channel and Super Channel sold advertising from offices in Japan, USA and Western European locations as well as from the UK. However no satellite television channel has yet generated adequate returns from advertising revenue. New transnational brands are slowly developing. 'Timotei' cosmetics were specifically developed as an international brand. The mainstays of transnational satellite advertising have been well established transnational brands such as Coca-Cola and Pepsi-Cola, Canon and Nikon, Philips *etc*. There are too few such brands to generate advertising revenues sufficient to cover costs.

Sepstrup (1985) in a study comparing a week's television advertising on Sky Channel and on the West German terrestrial broadcaster ZDF, observes that Sky Channel, in early 1984, had few brands advertised on its service, and that two product categories, 'chocolate, sweets, gum' and 'food products' accounted for 76 per cent of the commercial time sold by Sky Channel. Only 19 brands were advertised on Sky Channel in twelve weeks. These brands were insufficiently numerous (and audiences insufficiently large) to fund the programme services in which they were embedded.

However satellite television offered advertisers a mode of advertising hitherto unavailable in West European television; sponsorship. Sponsorship enables an advertiser to hitch a product to a particular programme and format and use the programme itself to build up a particular product image. In many respects, therefore, sponsorship is a more attractive advertising medium than spot advertising. Interestingly the BBC secured the US tyre manufacturer Goodyear as a sponsor for its weather programmes transmitted by Super Channel.

Even such sponsorship possibilities could not halt the 1988 retreat from the transnational market place by satellite television in Europe. Super Channel went bankrupt, and Sky

Channel decided to close its offices selling advertising time outside the UK, to re-orientate its re-launched service on ASTRA to anglophone audiences in the UK and Ireland, and to develop subscription funded services. The failure of the transnational market is but the most striking instance of the lack of economic success of advertising financed satellite television in Europe: no such satellite tv channel is yet reported to be profitable.

The channels that come closest to success are those which reach national television markets where there are restrictions on the supply of advertising time by terrestrial broadcasters and therefore an undersupply of advertising time at competitive rates. The most striking examples of satellite broadcasters advantaged by the undersupply of advertising on terrestrial television have been those transmitting to West German viewers, RTL Plus and Sat Eins. The main competitors to RTL Plus and Sat Eins, the three terrestrial services broadcast by the ARD and ZDF, consolidate television advertising into two blocks between 1800 hrs and 2000 hrs and do not transmit advertisements on Sundays or religious holidays. West German terrestrial television (from the point of view of the advertiser) undersupplies advertising time and packages it unattractively.

In other West European states – notably Scandinavia and Belgium – terrestrial television has been prohibited from carrying advertising. The ScanSat TV3 service (now transmitted from a satellite partly owned by Kinnevik, the progenitors of ScanSat) was established to offer advertisers access to the Scandinavian television audience and to exploit the opportunity presented by advertising free terrestrial television. However the window of opportunity identified by Kinnevik has been partially closed by the establishment of advertising funded terrestrial television in Norway and Denmark for the first time.

The satellite TV channels proposed for direct-to-home reception in the UK by BSB (British Satellite Broadcasting) and by Sky Television will depend, at least in part, on advertising revenues. The future UK terrestrial broadcasting environment, and therefore the extent and nature of the competition for viewers' attention and for advertising revenue, is unknown. A scenario offering significant changes for terrestrial television (and radio) is outlined in the UK Government's 1988 Broadcasting White Paper, 'Broadcasting in the '90s' but it is not yet clear how the government proposes to legislate. Neither is it clear how significant competition for advertising revenue from other media, such as print, will be; nor how healthy the UK economy, and therefore the demand for advertising, is likely to be during the 1990s. Estimation of possible advertising revenue for satellite television services orientated to the UK, therefore borders on the reckless.

The likelihood is, however, that the supply of terrestrial television advertising time in the UK will increase and that advertising revenue will as a result become a less secure source of funding for satellite television. For the government proposes an easing of sponsorship restrictions on terrestrial broadcasters, and an increase in the supply of advertising time on terrestrial television. Channel 5, if established as seems likely, will probably be funded, at least in part, from advertising. Channel 4 is likely to sell its own advertising and will therefore compete harder for advertising revenue than it now does, and there are to be three new advertising financed UK national radio channels. There will, therefore, be very stiff competition for advertising revenue and in this competition significant advantages will reside with terrestrial broadcasters.

It is possible that new advertising financed media (including an anticipated ten UK satellite television channels), will generate new money and that the advertising cake to be divided between them will be a larger one. But it may not be, and even if it is, it will probably not be large enough to cover the costs of all the proposed satellite channels. UK advertising spends are already the highest in Europe.

Advertising expenditure 1988 in $m

UK	11894
W Germany	9522
France	6843
Italy	4981
Spain	4407
Netherlands	2438
Switzerland	1985
Finland	1443
Sweden	1571
Belgium	981

Source: Advertising Association/James Capel. Cited *Sunday Times*, 12.3.1989, p. 13.

However there are some characteristics of the UK tv advertising market that offer hope to satellite television entrepreneurs. The existing market is relatively poorly supplied by the ITV companies, which currently have a monopoly on the sale of UK television advertising, selling time both on ITV and Channel 4.

The ITV/Channel 4 audience has declined to 51 per cent of the total audience and is older and poorer than the BBC audience (and is therefore less attractive to advertisers). Yet UK TV advertising is substantially more expensive than in other Western European countries, or in the United States or Japan.

Cost per 1000 adults for 30 seconds at peak rates 1986

France	£2.19
Japan	£2.37
USA	£2.45
Italy	£2.75
W Germany	£2.98
UK	£4.19

Source: Saatchi & Saatchi, Compton. *Advertising worldwide. European market and media facts*, 1988; cited *Financial Times*, 19.4.1988. p. 11.

The higher cost of UK television advertising relative to costs in other comparable markets may be due to an increasing demand for television advertising in the UK which has been unmet by a commensurate growth in supply of television advertising time.

MEAL (Media Expenditure Analysis Limited) instances the growth in UK, television advertising expenditure in the 1980s by important advertisers:

(a) By joint stock banks from £8.6 million in 1980 to £50.3 million in 1987.

(b) By insurance companies from £3.5 million in 1980 to £31.6 million in 1987.

(c) By building societies from £8.8 million in 1980 to £41.1 million in 1987.

Although the supply of UK television advertising time increased between 1980 and 1987 (as a consequence of establishment of Channel 4 and TV-am, and longer transmission times on commercial television) it seems clear that excess demand has pushed up the cost of television advertising time in the UK. Booz Allen & Hamilton (in *The Economics of Television Advertising*, 1988) estimate that the cost of television advertising rose by 57 per cent between 1980 and 1988. The Peacock Committee (1986) stated that NAR (Net Advertising Revenue) for ITV increased by 19.28 per cent between 1984–5 and 1985–6.

The growth in demand for advertising time, entry of new advertisers to the television advertising market, and growth in television advertising revenues suggest that there may be opportunities for satellite television to capture significant funding from advertisers. But the Peacock Committee also observed that TV advertising in the UK forms a much greater percentage of total advertising than in some major European countries (Peacock, 1986, p. 69).

Though Peacock did not consider the impact on the UK TV advertising market of satellite television, the Committee did consider the likely consequences of increase in the supply of television advertising were the BBC to sell advertising time. The committee noted: '... if anything the short-run effect of increasing slots would be to reduce total advertising revenue rather than to increase it'. (Peacock, 1986, p. 71)

Advertisers may support satellite television – if only to establish an alternative to the monopoly of the ITV companies. But their support is unlikely to be sufficient to ensure the financial viability of all, or any, of the projected UK satellite television channels. However predictions are extremely hazardous not least because the behaviour of all actors – advertisers, commercial television, and rival advertising media such as newspapers and periodicals – are unlikely to remain static in the changed circumstances presented by advertising financed satellite television.

High expenditure on UK television advertising may suggest that television is a more effective advertising medium in the UK (relative to alternatives such as newspapers) than in other countries, and that therefore division of television advertising revenues between terrestrial and satellite television is not a fixed sum game. Though the costs of UK television advertising are high it is possible that the service offered by UK television – in spite of the poor demographics of the ITV audience – is reasonably good. A restructured programme offer (like that already begun by the ITV companies with a revision of programme schedules involving changes such as the cancellation of *Crossroads*) may improve the demographics of the UK commercial television audience and thus make it even more attractive to advertisers.

However, if the UK advertising spend continues to grow, (as it is likely to do if the UK economy continues to grow), then satellite TV is likely to take a share of this growth and may capture some advertising expenditure from ITV. But even if there is a growing pool of advertising revenue potentially open to capture by satellite television the actual flow of revenue from advertisers to satellite broadcasters will depend on the ability of satellite television to deliver an audience to advertisers. It will be some years before satellite television can do this. BSB's business plan anticipates three years of losses before advertising revenues meet expenditure. This estimate looks optimistic. Doubtless ITV will improve the cost effectiveness of its advertising service in the meantime. The uncertain flows of advertising expenditure on satellite television will, moreover, be divided between several competing clusters of satellite services, offering, it seems, as many as ten channels of advertising financed television.

Some advertising finance for satellite television in the UK will doubtless be forthcoming. But the quantity and rate of its flow will depend on the rapidity of penetration of satellite TV receiving equipment, the size and demographics of the TV audience and the consumption patterns of the audience.

The flows of advertising revenue to services reaching a UK and Irish audience are likely to be stronger and faster than to services delivering a transnational audience. (Hence the shift in Sky Channel's business strategy from a European to a UK/Ireland market). For there are a plethora of goods and services already advertised and marketed in the UK and Eire, whereas there is a paucity of international brands. Whilst the whole transnational TV advertising market in Western Europe is estimated at £20 million, the UK TV market out-turn in the late 80s in contrast was approx £1.6 billion and may grow to £2 billion by 1995 (Coopers and Lybrand cited in *Cable & Satellite Europe*, Jan 1989, p. 37).

New Media Markets (vol. 7, no. 1, January 1989) estimates a first year, 1989, advertising revenue flow of £14.7 million for the three advertising financed channels distributed by Sky Television via Astra: Sky Channel £5.5 million, Eurosport £6.3 million, and Sky News £2.9 million. These revenues are unlikely to defray more than, at most, a fifth of running costs. It is likely that a positive cash flow for satellite television in the UK, let alone profitability, will be a long time coming.

Both the future audience share for satellite television and the future size of the UK TV advertising cake are imponderable. However AGB International's generous assumptions in respect of satellite television's share of television viewing and total UK television advertising revenue (based on an assumed satellite viewing share of 12 per cent and a television advertising revenue cake of £2.2 billion. as cited in *Financial Times*, 14.3.1989, p. 23) yield an estimated income for satellite television in the UK from advertising in 1993 to be £265 million. This revenue (of the order of one sixth of ITV/Channel 4 income in 1988) will be divided among advertising financed satellite channels, of which ten are proposed by the end of 1989. On average, therefore, satellite television might have revenues of about £26.5 million per channel: between a quarter and a fifth of the current revenue of Channel 4.

For advertising financed services such factors as audience demographics, audience size and number of hours watched, are crucial. Authoritative audience research is required to demonstrate to potential advertisers the size, demographics and consumption habits of the viewers exposed to advertisements. Such research is a significant cost element in the economics of satellite television. The established PETAR (Pan European Television Audience Research) satellite TV audience research costs its subscribers an estimated £1 million per annum, perhaps 5 per cent of gross revenues for the whole Western European satellite television market. These factors are unimportant, and this expenditure unnecessary, for the main alternative source of funding – subscription.

Subscription finance

Subscription funding for television raises revenues directly from consumers. It uses the price system as a means of signalling to suppliers the preferences of final consumers. For that reason it was recommended by the Peacock Committee as a funding mechanism for television in the UK.

Subscription funding of television in Western Europe is performed both via direct exchange between viewers and programmers and also via intermediaries (notably cable operators). In the latter instance, where there is no direct transfer of revenue (and thus, through price, information) from final consumers to programme suppliers, subscription funding offers few advantages over advertising finance in terms of its signalling between consumers and suppliers and in matching programme offer to audience demand.

Moreover the price of satellite television services to cable operators differs from the price charged for direct reception of satellite television. Some channels are free when received direct to home but are sold to cable operators, who then transmit the charges to final consumers. For example, the W H Smith channels, (which are free when received direct to home), Lifestyle, Screensport and Childrens' Channel, are sold to cable operators (for approximately £0.30, £0.60 and $0.35 per subscriber per month respectively). Such costs are transmitted to final consumers by the cable operators bundled in the monthly cable subscription charge. On the other hand premium film channels such as the projected Disney Channel which was planned to be bundled with Sky Movies will, when in operation, cost cable operators less per subscriber per month than is to be charged for direct-to-home reception (the proposed price for final consumption of Sky Movies/Disney Channel via Clyde Cable is anticipated to be £10.99 per month; the proposed price for direct-to-home reception £12 per month).

However subscription funding for television (whether terrestrial or satellite delivered) also entails significantly higher costs for service providers and consumers than does advertising financed television. Successful establishment of subscription television in France (Canal Plus) and the United States (Home Box Office) has encouraged advocates of subscription services in the UK. The differences between the broadcasting environments of France and the United States on the one hand and the UK on the other should not be underestimated. Nor should the failure of subscription finance in other markets (such as Canada) be forgotten. The undoubted success of subscription television in France (via terrestrial broadcasting), and the USA (via cable), is not necessarily a good guide to the future of subscription financed satellite television in the UK (or elsewhere).

The study of subscription television commissioned from Booz Allen & Hamilton by the Home Office (Home Office, 1987) is the most comprehensive treatment of subscription finance publicly available in the UK. However, it concerns subscription finance for terrestrial television and its findings are therefore of uncertain relevance to satellite television. It points out: 'Perhaps the most important of all the lessons to be learnt from overseas experience of pay television concerns the kind of programming that must be provided for the services to be viable'. (p. 28)

It finds that 'premium programming', *ie* feature films, high budget mini-series and sports are likely to best maximize subscription revenues. But there are limits to the extent to which subscription funding can sustain such services for 'There is no price level (and corresponding audience size) for which either BBC1 or BBC2 can recoup enough revenue from the market to finance the channel at the current expenditure level' (p. 126) and there is an interdependence between subscription and other services. Many viewers would choose to watch a free service in preference to one for which payment is required, and therefore the presence of advertiser supported channels ... is a major obstacle'. (p. 127)

No firm conclusions for the financial viability of satellite television subscription services can be drawn from this study of terrestrial television but the report notes: '... considerable consumer demand for extra television programming backed up by willingness to pay for it' and that: 'A category of programming which is particularly under-supplied in the UK is premium material such as high quality drama'. (p. 162)

There are, therefore, some positive indications for subscription funding of premium services. But the low take-up for the UK satellite-delivered subscription film channel Premiere, relative to comparable services in the USA and France, suggests that the consultants' positive findings concerning consumer demand should be regarded with caution.

In France the subscription service Canal Plus has undoubtedly been very successful. It began transmissions in November 1984, and by December 1986 had a positive post-tax income. By mid-1987 Canal Plus had 1.7 million subscribers – about 12 per cent of households in France – and was available off air via terrestrial broadcasting to 87 per cent of the French population. By the end of 1987 it had 2.1 million subscribers and a very low churn rate (92 per cent of subscriptions were renewed). In 1988 Canal Plus generated revenues equivalent to those of TF1, the most important of France's free-to-viewer terrestrial channels, and revenues in excess of those of each of the other significant terrestrial channels Antenne 2, FR3, M6, and La Cinq (La Cinq closed in early 1992). The importance of Canal Plus' subscription revenue is demonstrated by the fact that it commands only 3.1 per cent of French television advertising revenue, yet is as well funded as TF1, which commands 47.7 per cent of annual French television advertising revenue. Canal Plus anticipates a continuing growth in subscriptions and has already recovered its initial investments and start-up losses through a monthly subscription of 150 FF (and an initial decoder deposit of *ca* 400 FF).

The French broadcasting environment is different in several ways to that of the UK. VCR penetration in France is about half that of the UK, and the growth of the French VCR population was slowed by a variety of government measures, including a period when VCRs were required to be licenced. Canal Plus initiated night time transmissions in France (a dark period of only four hours in the week, and 24 hour programming at weekends) and transmits one new erotic film each month (repeated five times). Canal Plus' subscription programming is built around cinema films and is transmitted eight hours a day. It competes against five national networks where: 'Content is strictly controlled though "cahiers des charges", requirements to broadcast minimum amounts of certain types of content. The aim of this is one of public service ... . So the public service part of the French system is subject to considerable public control'. (Brynin, 1986, p. 4)

Canal Plus is thus significantly advantaged *vis-à-vis* terrestrial broadcasting in France (notably in its access to and freedom to screen feature films) and in comparison to what its UK equivalent is likely to be permitted to schedule. The Broadcasting Standards Council is unlikely to permit UK subscription channels to screen erotica such as *Deep Throat* (shown on Canal Plus), and UK broadcasting regulation is unlikely to tilt the 'playing field' so decisively against advertising funded broadcasters as has been done in France.

Canal Plus has successfully developed a mixed channel with the majority of its transmissions financed by advertising and therefore available free to viewers. These viewers, watching free (or at most having had to purchase a new receiving antenna) are exposed to promotional trailers of programmes to be transmitted in scrambled form and available only to subscribers. Canal Plus has, therefore, a cheap and effective system of marketing its product to non subscribers. Because Canal Plus' scrambling system is simple – which incidentally makes it vulnerable to piracy – decoder costs are low, about £65. Canal Plus has benefitted from cheap terrestrial transmission and reception and from the underdevelopment of entertaining programming on the established French networks as well as from its own innovatory and successful mix of programming and methods of finance. Another advantage it has is that its hardware costs are lower than is anticipated for UK satellite television since the widely publicized 'under £200' for Amstrad's receiving antenna and electronics for services transmitted from the Astra satellite did not include installation or the cost of decoders for scrambled subscription services. (Costs of hardware in the UK have declined significantly since 1989 – see preface.) Canal Plus has demonstrated the possibility of successfully establishing a subscription television service built around premium programming.

However, it is unlikely that this pattern could be replicated in the UK. Canal Plus' erotic films would be unacceptable to UK regulators. Further, its competitive advantage, *vis-à-vis* other French television channels, of being permitted to show new films on release (other channels must wait three years, or two years if a co-producing participant in production) and on Wednesday, Friday and Sunday evenings when other channels are prohibited from screening films, is not likely to be copied in the UK. Canal Plus is also vulnerable to its own success: once subscriptions exceed 2.9–3 million its liability to rights payments rises sharply.

There are European experiences of subscription funded television other than Canal Plus from which tentative conclusions about the viability of UK subscription services can be drawn. The satellite delivered subscription film channel, Esselte (which has viewers in Scandinavia and the Low Countries), was the first West European satellite television channel to achieve a positive cash flow. It estimates that with 350,000 subscribers (10 per cent of Swedish cable homes and 5 per cent of Dutch and Belgian cable homes) achieved in early 1989, it now breaks even. However Esselte has yet to generate revenues sufficient to recoup its accumulated losses. In 1988 Esselte lost 60 million Swedish Kroner on an annual turnover of 500 million Skr.

In the UK, the satellite-to-cable (and about 2500 TVROs) film channel Premiere claimed in 1988 to have a positive cash flow on a subscriber base of *ca* 106,000 in 1988, with a growth in subscriptions in excess of 100 per week. Its subscriber base had continued to grow throughout the channel's history; in 1986 by 100 per cent and in 1987 by 50 per cent. Premiere's signals are customarily transmitted scrambled, though cable operators will from time to time de-scramble the Premiere signal for subscribers as a promotional device to augment subscriptions.

Like all subscription channels Premiere's revenue depended on maintaining subscriptions rather than maximising viewing of its service. Even so, it has secured higher ratings among its viewing population than either BBC2 or Channel 4. However its subscriber base is perhaps 10 per cent of its potential; the UK homes passed by cable. As a monopoly premium film channel (prior to the inception of transmissions of Sky Movies from Astra) Premiere covered its costs. It was able to acquire feature films on a licensing basis which provided for 10 screenings a year and with the fee linked to the number of Premiere subscribers. However the future of Premiere is highly uncertain. The channel declined to match the prices paid by Sky Movies and BSB for film properties and closed when it ran out of product and Sky Movies and BSB Screen came on stream.

The Home Office Subscription Television study comments on the unsatisfied UK demand for premium programming. An important reason for UK interest in Canal Plus and Home Box Office is their status as precedents and their demonstration that the supply of premium programming can be increased by recycling the revenue flows from subscription into programme production. Certainly if the unsatisfied demand for premium programming is to be met then new production, funded by 'new money' from subscriptions will be required. A representative of Premiere (interviewed by the author in June 1988) commented on the paucity of attractive film properties and the necessity to programme films which had not previously been shown theatrically. Films which had not previously been shown in cinemas were usually inferior (that was why they had not been screened) and in any case had not previously been promoted for theatrical exhibition and were therefore unlikely to attract viewers. Premiere's programming strategy was to screen one blockbuster film a month, to support it with good quality previously released properties and to fill the schedules with 12–14 unknown and often unreleased properties. All films screened on Premiere (and its sister tape-to-cable channel HVC) have to be BBFVC-certificated. The channel is therefore constrained in its ability to supply the 'unsatisfied demand for sex and violence' among viewers (interview with Premiere, June 1988). The capacity of other channels to satisfy this demand will be similarly constrained, by the activities of the recently established UK Broadcasting Standards Council.

Film channels do not compete only against each other for product (the UK experience of two satellite film programme streams bidding against each other for product, and the elimination of a third, may not be replicated in other European markets) but also against other distribution media such as terrestrial television and video cassette retailers and renters. If 40,000 cassettes of a premium title are sold the UK pre-recorded cassette market may return £0.5 million to a film's producer. The video market is, therefore, a powerful competitor for preferential access to rights for premium film titles. It is likely that the existing hierarchy of exploitation for film titles will be maintained, that is theatrical exhibition, video cassette sale/rental, premium cable/satellite, terrestrial broadcast television. Thus assessment of the likely future development of satellite television requires consideration of other media of distribution of premium material. Subscription film channels distributed by satellite, though likely to enjoy earlier access to new films than does terrestrial television, or advertising financed satellite television, will remain third in line after the video rental market.

As the Home Office study states; there is undoubtedly *demand* for more high quality drama by UK television viewers. However, the study does not recognize that this unmet demand is related to an international undersupply of suitable product. This undersupply has constrained the programming and development of existing premium subscription services and will constrain future services.

The premium material for which unsatisfied demand exists is essentially programming – whether feature films or teledramas – of the type produced for markets where suppliers have historically enjoyed quasi-monopolistic access to audiences and revenue sources. The creation of a more competitive, multi-channel broadcast market in which more channels (and consequently more hours to be programmed) share a revenue pool of uncertain growth, decreases the possibility of financing such premium programming. A fixed, or at best slowly growing sum of finance, will have to fund an enormous quantitative growth in hours of programming, which will mean revenues spread wider but thinner. In consequence premium programming is likely to become less, rather than more, affordable.

Subscription funding of channels also adds transaction costs to the cost structure of television (whether satellite, cable or terrestrially broadcast). Not all the 'new money' from subscription will go 'on the screen' (or into returns to investors). Billing subscribers, distributing decoders and/or smart cards, scrambling the signals transmitted, policing signal piracy – all these activities will inflate the channel's cost structure. (A 'smart card' is a card with machine-readable information encoded upon it. Such information may simply be encoded on a magnetic stripe – like a credit card – or the 'smart card' may contain a microprocessor.) There will undoubtedly be some countervailing benefits for those offering subscription funded programme streams, such as access to data concerning final consumer behaviour, but these are unlikely to outweigh the substantial transaction costs involved in running, and policing, a subscription funded system.

Channel operators offering subscription funded services in direct-to-home (DBS) mode are likely to enjoy significantly better flows of consumer information than have operators (such as Premiere) of established satellite-to-cable services. This derives from their having only one type of customer to whom the channel requires to be marketed. Existing satellite-to-cable services delivered by low powered satellites (such as Premiere) require marketing and promotion, to cable operators as well as to final consumers. Moreover, information from the subscriber base often stops with the cable operator rather than being relayed to the service provider.

It seems that neither advertising finance, nor subscription finance offer assured flows of revenue to satellite television programmers in quantities sufficient to cover their costs of doing business (whether addressing a transnational or UK national audience). The financial prospects for satellite television in the United Kingdom are very uncertain indeed.

4 The audience

The audience for satellite television is best considered under two heads: viewers' access to satellite television and the behaviour of viewers who have that access. The discussion that follows of these two, related, questions draws on the findings of audience research conducted during the 'second generation' of satellite television in Western Europe. Little is known about the audience's relation to 'third generation' DBS television which, at the time of writing, had been established for less than three months.

Access

The most important channel of access to satellite television is via cable. However, access to satellite television via cable cannot be assessed simply by establishing the number of cable connections in a particular population of potential viewers. For cable systems differ in their capacity to carry satellite signals: some venerable UK cable networks carry only one satellite channel. And access to satellite television does not correlate with consumption of it. At least three distinct factors are therefore relevant to consideration of consumption of satellite television.

The first of these is access to signals. For second generation satellite television the crucial factor was availability of cable, because few viewers accessed signals via an individual dish (TVRO). In respect of the UK, before the inception of direct-to-home satellite television service from Astra, the UK Department of Trade and Industry (DTI) issued (1988) about 5000 licences for individual TVROs in the UK. Now, after commencement of a DBS service (designed to make satellite television available to viewers at affordable cost without the intermediary of cable) little seems to have changed. Although there is no consensus about the rate of growth of the UK TVRO population, consultants' estimates, (to be preferred to those of directly interested parties such as dish manufacturers and satellite television programmers), suggest a low dish population in the UK and Western Europe for the foreseeable future. (Logica estimate a UK dish population by 1994 of 600,000; CIT, a European population of 1m by 1996.) So, it seems that even in the third generation of satellite television, cable will continue to be a very important intervening variable between viewers and service providers, and that therefore the fate of satellite television will continue to depend on the fate of the cable industry.

Second, the nature of the cable system through which viewers access satellite television is an important factor. Of the UK homes which subscribe to cable only about a quarter are connected to broadband multi-channel networks. Many UK cable systems have a very limited capacity to relay satellite television signals. In addition, cable operators are bound by a requirement to relay UK-authorized broadcast services. This has meant that terrestrial television has taken priority over cable.

Third, access to satellite television (via cable TVRO or otherwise) is only a *precondition* of consumption. Not all potential viewers are actual viewers. Even in the US television market, distinguished by extensive cabling penetrating more than half US households and a viewer population familiar with a multi-channel viewing environment, viewers customarily access only a few channels even when they could choose to access many. In cable homes in the United States only an average 8.2 channels are accessed (*Screen Digest*, January, 1989, p. 8) and the three major terrestrial television networks (ABC, CBS, NBC) account for more than 70 per cent of the television viewing of US audiences.

A survey of Amsterdam cable subscribers (*Cable & Satellite Europe*, 3.1988, p. 57), albeit of only 389 respondents and therefore of uncertain reliability, demonstrated that the Dutch viewers principally demand terrestrial services (both Dutch national and neighbour country signals) rather than satellite television. In answer to the question 'What channels are you interested in watching?', the sample gave the following replies:

	% very interested
ARD (German terrestrial)	31.9
Arts (UK satellite)	9.3
BBC1 (UK terrestrial)	45.8
BBC2 (UK terrestrial)	40.6
BRT1 (Belgian terrestrial)	29.8
BRT2 (Belgian terrestrial)	37.8
FilmNet (Swedish/Dutch satellite)	8.0
NOS1&2 (Dutch terrestrial)	81.5
Sky (UK satellite)	26.2
Super Channel (UK satellite)	21.1
TV5 (French satellite)	7.5
WDR (German terrestrial)	25.2
ZDF (German terrestrial)	29.8
Kabel-nieuws (Dutch cable)	11.8
Lokale TV (Dutch cable)	17.2
Parool TV (Dutch cable)	19.0

Cable & Satellite Europe, 3.1988, p. 57.

The best estimates of access to satellite TV via cable are those published quarterly in *Cable & Satellite Europe*. They establish Europe-wide league tables of possible (rather than actual) consumption as follows:

	February 1988	November 1988
Sky Channel	10,869,773	13,874,853
Super Channel	9,660,000	13,390,000
TV5	7,029,685	9,410,767
RAI	3,575,000	3,575,000
Worldnet	3,452,212	Ceased transmission
Drei Sat	3,432,280	4,920,000
Sat Eins	2,873,280	4,173,000
RTL Plus	2,868,280	4,400,000
Arts Channel	Not recorded	13,611,526

All the above channels are carried on the Eutelsat F1 13 °E satellite, but the connections stated for Sat Eins and RTL Plus undoubtedly understate access to these channels since significant numbers of West German viewers receive these signals via terrestrial broadcasting or re-broadcasting.

The Eutelsat F1 13 °E 'first division' is followed by services transmitted from the Intelsat VAF12 60 °E satellite. The following German language services were all accessible to 2,568,280 cable households in West Germany via Intelsat VAF12 in February 1988: BR3, Eins Plus, Eureka, Tele 5, WDR 3. By November 1988 all were accessible to at least 3,276,000 cable homes and some to more (to a maximum of 3,609,000). The third most important bundle of services are those carried on the Intelsat VAF11 27.5 West satellite.

The French language service transmitted from the Telecom 1B satellite was available to the following number of cable households:

	February 1988	November 1988
MTV	2,212,425	4,103,297
Childrens' Channel	887,000	1,160,000
Screensport	241,110	1,015,530
CNN	233,209	849,412
Lifestyle	203,300	835,980
Canal J	40,020	Not recorded

The channels itemized above are all either 'free' or low-pay. Channels which charge premium rates for access, such as the three film channels listed below, are not as attractive to cable networks as are free and low-pay channels and are therefore accessible to relatively fewer cable subscribers than are the free/low-pay channels.

	February 1988	November 1988
FilmNet (on Eutelsat F1 13 °E)	106,000	213,000
Premiere (on Intelsat 27.5 °W)	80,000	103,000
Teleclub (Eutelsat F1 13 °E)	43,700	64,000

Comparison of these February and November 'snapshots' of potential viewership of satellite television via cable in 1988 is revealing. First, it is important to note how much the figures change in a short period: the changes that refer to the Arts Channel and to Worldnet are particularly striking. The Arts Channel's successful conclusion of an agreement to share a transponder with Sky Channel (Arts Channel being transmitted during night hours) lifted viewer access to Arts Channel from a level which was too low to be detected in survey research (the channel was about to cease transmission) to the level of being one of the channels most widely available. In contrast Worldnet ceased transmission following a policy decision by the United States Congress, the source of its funding. Rises in levels of accessibility for other channels followed different events: for German language channels, such as Eins Plus and Tele 5, the extension of new-build cable networks in West Germany was important. Other channels, such as Childrens' Channel, extended their potential viewership by succeeding in gaining access to cable networks from which they had hitherto been excluded.

As comparison of the cases of FilmNet (reputedly the only satellite channel in Europe to achieve a positive cash flow) and Sky Channel (which scored a pre-tax loss of £8,452,544 in 1987–8) demonstrate, access to homes is not directly transferable to the company's financial bottom line. The most widely available subscription funded channel, FilmNet, approaches profitability whilst an advertising-financed channel available to 65 times more homes makes substantial losses (albeit reduced from the previous year).

In March 1988 Sky Channel, the channel which has consistently enjoyed higher access to viewers' homes via cable than any other satellite TV channel, had 30.5 million potential viewers. For much of its history Sky has scrambled its signal to inhibit non-cable reception and therefore cable access to Sky Channel is likely to be a good proxy for total access.

Impressive though Sky Channel's access to homes was, access does not correlate closely with actual viewing. Viewing of Sky Channel (and indeed other programme streams such as Super Channel) was not commensurate with the access to homes achieved by the channel. In spite of greater success than any other channel in maximising access to viewers across Europe, in 1988 Sky Channel was unable to realize this achievement in profitable trading. It re-oriented its corporate strategy away from attempting to establish itself as a transnational advertising funded channel, towards becoming a UK (and Ireland) channel funded by a mixture of advertising and subscription funding. Sky Channel's new strategy emphasizes three factors previously absent: an anglophone audience, direct-to-home reception, and subscription funding. Though there is a continuity in name and in some personnel and programming, Sky Channel is quite a different entity post February-1989 (the inception of transmissions from Astra) to what it was prior to that date.

Access to Sky Channel

	Country networks (Cable & SMATV)	Households
Netherlands	508	3,379,650
W Germany	1039	3,135,040
Switzerland	209	1,396,920
Belgium	58	1,103,322
Sweden	302	420,677
Denmark	275	390,669
Finland	168	361,471
Norway	256	338,789
Austria	155	325,661
Ireland	22	314,177
UK	111	263,341
Hungary	23	104,694
Luxembourg	25	85,713
France	68	78,960
Spain	348	72,843
Portugal	72	28,072
Greece	5	1,017
Iceland	21	690
Yugoslavia	1	420
19 Countries	3666	1,802,126

= 30.5 million potential viewers

Source: Sky Channel, 1988.

Sky estimates its penetration of different national television markets (January 88)

Households	Country TV % of all households ('000)	SKY households ('000)	AS TV
Netherlands	4345	3360	63
W Germany	25,336	2795	11
Switzerland	2494	1288	52
Belgium	3500	1100	31
Denmark	2200	375	31
Sweden	3333	374	11
Norway	1550	316	20
Finland	1800	301	17
Austria	2780	310	11
Ireland	918	296	32
UK	20,596	256	1
Luxembourg	118	84	71

Viewing

The best evidence on satellite TV viewing comes from the PETAR (Pan European Television Advertising Research) surveys. PETAR is funded jointly by Sky Channel, Super Channel, Screensport and Lifestyle (W H Smith Television), Sat Eins, RTL Plus, RAI, MTV, CNN, McDonalds and the IBA. The annual cost of the PETAR survey research is about £1 million: a substantial sum for loss making businesses. But authoritative audience research is necessary in order to demonstrate to advertisers the utility of satellite television advertising.

The PETAR survey of Spring 1987 was the first authoritative study of consumption of satellite television in Europe. It, and the survey of the following year, was based on diary research. In 1987, 2651 respondents completed four-week diaries. In 1988 there were 4807 respondents. The samples were drawn from individuals aged four and above able to receive at least one channel of satellite television via cable in Austria, Belgium, Denmark, Finland, Ireland, Norway, Sweden, Switzerland, West Germany and the UK. Among PETAR findings were the following:

% share of viewing by individuals aged 4 years and above of satellite television in homes with access to satellite television via cable in 1987 and 1988 by country and total

		Nordic	Belgium	Switzerland	Netherlands	W. Germany	Total
RAI	87	–	2	4	–	–	1
	88	–	1	5	–	–	1
Sky	87	20	1	2	6	2	5
	88	11	1	3	4	–	4
Super	87	7	–	1	2	1	2
	88	5	–	1	1	–	1
Sat 1	87	–	–	–	–	14	4
	88	–	–	2	–	20	6
RTL+	87	–	–	–	–	13	4
	88	–	–	–	–	10	3
All Sat	87	27	3	7	8	30	16
	88	24	3	10	6	32	16

NB 1987 and 1988 findings are not precisely comparable
The 1987 study comprehended 'Scandinavia', and the 1988 'Nordic Countries'.
The 1987 study considered satellite television, the 1988 study commercial satellite television.
Percentage shares of less than 1 per cent have been shown as zero.
Source PETAR surveys 1987, 1988 from *Cable & Satellite Europe*, 11.1987 and 11.1988, and Sky Channel press pack.

The 1988 PETAR survey gave the following viewing share percentages for the UK and the Republic of Ireland. No comparable statistics were available in 1987.

	All commercial satellite	Sky Channel	Super Channel
UK 1988	26	14	2
Ireland 1988	19	13	3

Clearly there is no unitary European experience of satellite television. Overall, in a year in which access to satellite television doubled (from about 5 per cent to 10 per cent of West European television households), its share of viewing time remained remarkably constant at 16 per cent. However, Sky Channel experienced a *reduction* in its overall European share of viewing and a striking *decline* in its share of viewing in Scandinavia. Super Channel and RTL Plus suffered similarly. On the other hand Sat Eins, the rival German language commercial channel achieved an impressive *increase* in its viewing share in German language markets.

The phenomenon of satellite television, which varies from country to country and between different channels needs to be disaggregated by age of viewer as well as by country and channel. Younger viewers watch proportionally more satellite television than do older viewers. In 1988, Europe-wide, children's viewing of satellite television accounted for 26 per cent of their total television viewing, as opposed to a 15 per cent share of adult's viewing. In the UK satellite television accounted for 41 per cent of childrens' viewing and 22 per cent of adult's viewing; in West Germany for 47 per cent of children's and 31 per cent of adult's viewing (*Cable & Satellite Europe*, November 1988, p. 32).

The bias of young people's and children's viewing towards satellite television echoes a general European and North American phenomenon. Consumption of own-country television is correlated with age: younger viewers have a stronger disposition to watch exogenous television than do older viewers. This disposition is doubtless amplified in the West European television environment by the fact that satellite television offers channels tailored to different age strata in the audience. MTV, Children's Channel, Kindernet are all tailored to the youth and child age groups. Sky Channel also scores well with younger viewers, doubtless a response to ,inter alia, its Dutch-produced DJ Kat show orientated to children. Whilst satellite television (in particular Sky Channel) achieved a reasonable reach within the universe of cable homes in 1987 what is most striking overall is how *little* satellite television was watched.

There does not appear to be a consistent international relationship between access to, and consumption of, satellite television as the following examination of four cases shows. The cases considered (Sweden, The Netherlands, West Germany and the UK) account for 60.3 per cent of West European cable homes (Sweden 3.5 per cent, The Netherlands 28.2 per cent, West Germany 26.5 per cent, UK 2.1 per cent: Source *Cable & Satellite Europe*, 11.88, p. 32).

Sweden

Sweden combines *low access* to satellite television (8 per cent of homes according to a Sveriges Radio estimate of 1988, or 11 per cent according to a Sky Channel estimate of the same year) with *high consumption* by viewers with access to it. Satellite television is variously estimated to account for a 30 per cent share of Swedish viewing (Sveriges Radio estimate), or a 27 per cent share of viewing (Sky Channel estimate). Swedish viewers use terrestrial television relatively little in comparison to viewers elsewhere (an average of 13.76 hours per week, according to Sveriges Radio, or 13.4 hours per week according to Sky) whereas Swedish viewers of satellite television watch for 16.2 hours per week; that is nearly 21 per cent more than do terrestrial television viewers.

The low overall consumption of television and relatively high consumption of satellite television in Sweden suggests dissatisfaction with terrestrial television and that satellite television provides a welcome additional service.

The Netherlands

The Netherlands combines *high access* to satellite television (50 per cent according to an NOS estimate in Bekkers, 1987; 63 per cent according to Sky) with *low consumption* (NOS, 1986, estimate 6 per cent; Sky, 1986, estimate 8 per cent) by viewers with access it. Dutch viewers use terrestrial television for 16.9 hours per week (Sky estimate) but their consumption of satellite television *declined* between 1985 and 1988.

> Terrestrial services share of Netherlands television viewing: 1985 78 per cent, 1986 84 per cent, 1988 93 per cent.
> Satellite TV share of Netherlands viewing: 1985: 10 per cent; 1986: 7 per cent; 1988: 6 per cent.
> (Sources: Bekkers, 1987; PETAR, 1988, cited in *Cable & Satellite Europe*, 11.1988, p. 32).

West Germany

West Germany combines *moderate access* to satellite television with *high consumption* of it. 34.4 per cent of German households were (1987) passed by cable, of which 36.1 per cent subscribed to it, a penetration of about 14 per cent (DBP Press Release 29.1.1988). Sky Channel was accessible to 11 per cent of TV households, and RTL Plus and Sat Eins to all cabled households, that is to 14 per cent of West German television households. RTL Plus and Sat Eins also have a significant reach via terrestial transmission. These are re-broadcast from terrestrial transmitters employing satellite feeds. Viewers receiving signals via terrestrial rebroadcasting accounted for (1987) an additional 770,000 households for Sat Eins and an additional 790,000 households for RTL Plus. Access to television homes in West Germany by RTL Plus is further increased via reception of signals from neighbour country terrestrial broadcasting (RTL Plus has an additional 450,000 households in West Germany receiving terrestrial broadcast signals from its Luxembourg transmitter). With the addition of households receiving signals via terrestrial broadcast, or re-broadcast, the penetration of Sat Eins and RTL Plus amounted in 1987 to approximately 16 per cent of television households in West Germany.

Viewing of satellite television in West German cabled households in 1987 (*Media Perspektiven*, 1/1988) was for approximately 47 minutes per day. That is, satellite television accounted for a 29 per cent share of viewing. PETAR, 1988 (based on diaries for April/May 1988) estimated satellite television's share of West German viewing in households with access to satellite television to have increased to 32 per cent. Increased access to satellite television increased overall consumption of television by 9 min per day, which is approximately 6 per cent. It reduced consumption of German terrestrial services by 36 min per day in 1987. In other words, cable households watched only 25 per cent less terrestrial television than non-cable households. The preponderance of West German audiences' viewing of satellite television was German language services.

In 1987 satellite TV viewing per day by West German cable viewers amounted to 62 min of which:

15 minutes	RTL Plus
22 minutes	Sat Eins
4 minutes	Drei Sat
2 minutes	Eins Plus
1 minute	Sky Channel

These statistics suggest that West German terrestrial services undersupply the type of programming desired by audiences, and that the provision of satellite television has not only displaced consumption of terrestrial services but also increased consumption of television overall. And that German language satellite services provide a service preferred by viewers to that provided by Sky Channel. It is striking how German viewers' consumption of Drei Sat and Eins Plus (satellite channels programmed by German terrestrial public service broadcasters with repeats of terrestrially broadcast programmes) exceeds that of Sky Channel and Super Channel, even though these offered programmes that were both new to German viewers and more entertaining than those of Drei Sat and Eins Plus. The attractions of novelty and entertainment did not outweigh the disadvantage of the English language content of Sky and Super Channels' programming. The 1988 PETAR survey shows a *growth* in satellite television's share of viewing in West German cable homes, and also in the number of homes connected to cable. The total impact of satellite television on West German viewers increased as a product of these two factors (with Sat Eins the main beneficiary of this growth). At the same time there was a striking decline in the share of viewing of Anglophone services.

Two factors seem to have structured West German consumption of satellite television: first a desire for programming undersupplied by terrestrial broadcasters – principally feature films and television series – and second a desire for programming in German.

The United Kingdom

The UK exhibits *low access* to satellite television and *low consumption* of it. The UK has 1.51m homes passed by cable and 272,559 homes connected to cable networks, a penetration of 18.1 per cent of homes passed (homes passed by new-build broadband networks have an even lower penetration ratio). Most worrying for cable operators has been the slowness of the rise in the rate of penetration of cable. In fact, at some points it has even declined. In 1986 cable penetration in the UK was 13.1 per cent of homes passed. In January 1988 penetration rose to 18.3 per cent but *declined* to 18.1 per cent in January 1989. In new-build broadband areas penetration rose from 11.4 per cent in 1986 to 14.7 per cent in January 1988, but *declined* through 1988 (to 14 per cent in October) to revert to 14.7 per cent penetration in January 1989. Most UK cable households are connected to narrowband networks which have limited capacity for satellite television (some systems have capacity for only one satellite service). A minority of UK cable subscribers are connected to broadband systems; they number about 63,000 (all statistics from *NMM*, 15.2.1989, p. 7).

Penetration via cable of satellite television in the UK

Channel	1987	1988	1989
Sky Channel	248,169	261,864	270,747
Super Channel	124,194	148,458	166,070
Children's Channel	132,573	158,127	158,070
Screensport	113,989	115,097	116,630
MTV	97,169	106,033	105,448
Lifestyle	115,153	68,631	78,326

Sources: *Cablegram*, 2.1988, p. 2; *New Media Markets*, 15.2.1989, p. 7.

In 1987 Sky Channel achieved an 11.3 per cent share of viewing in Sky Channel homes (a phrase virtually synonymous with cable homes) and Super Channel 3.8 per cent in its universe. Satellite TV channels in the universe of UK cable homes achieved altogether a 28.5 per cent viewing share.

% share of viewing all individuals aged 2+ in all cable homes

ITV	36.7	
BBC1	25.1	
Premiere	8.8	
Sky	7.0	
Channel 4	4.8	
BBC2	4.4	
Children's Channel	4.2	
Screensport	2.5	
MTV	2.4	
SuperChannel	1.5	
Lifestyle	1.2	
TV5	0.4	
RAI 1	0.2	
Arts Channel	0.2	
CNN	0.1	
All Terrestrial	71.0	
All Satellite	28.5	
Non satellite cable	1.5	(does not add up to 100 %)

Based on 2-week diaries of 770 respondents for the first two weeks of December 1987; Source: *NMM*, 6.4, p. 7.

Though satellite television achieved an impressive share of viewing in cable homes (and both Premiere and Sky achieved higher shares than Channel 4 or BBC2) these achievements should be relativized by recognising that cable homes are likely to be those disposed towards consumption of satellite TV. 85.3 per cent of homes passed by cable did not subscribe to cable. Moreover cabled areas are likely to be those judged most likely to yield high penetration: that is, cabled areas are likely to be *more* receptive to cable/satellite television than is the UK television population as a whole. And subscribers to cable are likely to be more receptive to cable/satellite services than those passed by, but not subscribing to, cable. But even within the UK cable/satellite population of more than a quarter of a million TV households (many of which will be able to receive only one satellite TV channel) consumption of cable and satellite television is *declining*. A 26 per cent share of viewing in 1988 represents a fall from a 30 per cent share of viewing in 1987 (PETAR surveys) and a further decline from the 46 per cent for cable and satellite channels within the same universe in 1986 (Saatchi & Saatchi Compton, p. 6, 1987). The UK experience of satellite television accessed via cable has not been one to encourage satellite television programmers, cable operators or advocates of the 'New Media'.

Programming

The technological potentiality for satellites to deliver alternatives to national terrestrial television, and to create new transnational forms of television, now exists. There are important intervening variables (such as transmission standards, price charged, advertising revenue *etc*) which significantly affect the success or failure of satellite television. However, the single most important factor, and the *sine qua non* of success, is delivery of programming attractive to the final consumer. The attractiveness of programming is not an absolute, but is related to the available alternatives, and to the needs and desires of audiences. Successful programming of satellite television demands that a number of criteria be satisfied.

First, in a multi-channel television environment, the environment that satellite television is creating and/or amplifying, channels require to be strongly 'branded'. They need a distinct identity, so that viewers will seek them out for a particular desired viewing experience. The brand can come from a particular kind of programming, such as sport, news, or feature films. This rationale has been followed by 'thematic' channels such as 'Screensport' or 'Children's Channel'. Or the brand can come from a style. For example Sky Channel, though offering a mixed programme schedule, has a clearly demotic, youthful, fun image. It has successfully established its brand. Other channels – *eg* Super Channel – have not yet established a distinct brand and offer neither a particular type of programme nor a particular distinctive style.

Second, the programme schedule for satellite television requires to be regular, easily comprehensible and predictable. Audiences need to know that at time x on date y on channel z a particular programme or programme type is scheduled: 'on Thursdays at 7 pm on Channel 13 it's M*A*S*H'. The successful satellite television schedule will therefore be organized around consistent and regular programme junctions (*eg* on the hour) and will look more like the 'stripped' schedules of United States television (an adaptation to the US multi-channel environment) than the present UK or West German terrestrial television schedules where programmes of different lengths and type are shown, and which have few consistent programme junctions or patterns. The search time to find attractive programming in a multi-channel environment may be unacceptably high if channels are not strongly branded and programme junctions regular. There is therefore an important competitive advantage for channels with a strong brand which schedule programmes in simple and recurrent patterns. Channels seeking a multi-national audience (and therefore scheduling programming not in the native language of the whole audience) should avoid programme material with a high speech content – and in particular speech not uttered straight to camera – which is subject to misunderstanding by audiences and is therefore unpopular with viewers. As Super Channel discovered to its cost, UK television dramas such as *Rumpole of the Bailey* which have characters' interior monologues on the soundtrack with no evident linkage to the images, and documentaries with voice over narration unmotivated by the images shown: such programmes are not well adapted to an audience of non-native speakers of English. For reasons such as these much UK 'quality' programming is unsuitable for a transnational audience; in contrast much US programming has the desired qualities that make for international comprehensibility.

Cultural screens differentiating European TV audiences

There are important 'cultural screens' that differentiate Western European audiences from each other. As will be seen, there is scant unity of taste and preference among the West European audiences for satellite television. Hence satellite television channels which have attempted to construct a transnational audience have experienced intractable problems in devising a schedule that appeals sufficiently to a cross-European audience. Super Channel's attempts to attract European audiences with *The Best of British* – 'quality' UK television dramas such as *Harry's Game*, *Nicholas Nickleby* and *Brideshead Revisited* – were not successful. Other UK dramas were perceived, by West European television viewers, as being 'too violent' and 'too realistic'. Police series such as *Taggart* were anathematized by European viewers in comparison to US series such as *The A Team*, because, although the level of violence in UK programmes is lower (punches and kicks rather than shootings), it is shown *more realistically* than on US television. UK tv news has similarly been perceived as too violent, and the studio-based drama characteristic of British television was found to be unacceptable in some European markets. UK audiences have developed a tolerance for studio sets that other European audiences, more habituated to location based film dramas have not (interview with PETAR, May 1988). More positively, for UK programme producers, there is a pervasive West European demand for feature films and for UK comedy of the Benny Hill, Kenny Everett kind.

The varieties of preferences of West European television viewers are signalled in the findings of PETAR's research into reception of the UK satellite television channel Super Channel. Super Channel began its life as a programme stream designed to draw on UK programming and capture the attention of West European audiences with a schedule of *The Best of British Television*.

The owners of Super Channel, the UK ITV companies hoped to use satellite television as a means of directly distributing their programmes to European viewers, rather than having to use as intermediaries the terrestrial broadcasters to whom programmes had formerly been sold.

1987 PETAR audience research on Super Channel programming

PETAR identified *Benny Hill* as the most popular of Super Channel's programmes in Spring 1987 (ironically *The Benny Hill Show* is a production of Thames Television – the sole ITV company that declined to participate in Super Channel and lost its London franchise in the recent round of applications). Of a potential audience of about 16m viewers for the Sunday 7.30–8.30 pm screening, *The Benny Hill Show* reached a peak 2 per cent. However, between individual national audiences, and between men and women, response to the most popular programmes differed significantly.

Netherlands Men 16+		Average shares Women 16+	
1. *Benny Hill*	2.9%	*Benny Hill*	1.8%
2. *Kenny Everett*	1.8%	*Mistrals Daughter* (US mini series)	1.3%
3. *Survival*	1.6%	*Kenny Everett*	1.3%
4. *Super Sport*	1.4%	*Armchair Adventure*	1.2%
West Germany Men 16+		Average shares Women 16+	
1. *Spitting Image*	2.1%	*Life on Earth*	0.8%
2. *News*	0.9%	*Wild World*	0.8%
3. *News*	0.7%	*Shirley Bassey*	0.7%
4. *News*	0.7%	*Great Railway Journeys*	0.7%
Scandinavia Men 16+		Average Shares Women 16+	
1. *Benny Hill*	6.6%	*Benny Hill*	5.6%
2. *Benny Hill*	5.3%	*Princess Daisy*	5.4%
3. *Kenny Everett*	4.7%	*Mistral's Daughter*	5.2%
4. *Princess Daisy*	4.1%	*Limestreet*	3.7%
Belgium Men 16+		Average Shares Women 16+	
1. *Super Sport*	1.8%	*News*	1.0%
2. *Super Sport*	1.8%	*Princess Daisy*	1.0%
3. *Super Sport*	1.8%		
4. *Princess Daisy*	0.9%		
Switzerland Men 16+		Average Shares Women 16+	
1. *European Top 40*	1.8%	*European Top 40*	0.9%
2. *Countdown*	0.9%		
3. *Living Body*	1.1%		
4. *Busman's Holiday*	0.9%		

Source: PETAR Survey, March–April 1987.

The difficulty of programming a satellite television channel to maximize viewing among West European audiences is graphically exemplified by this response of audiences to Super Channel's programme offer. There appears to be scant communality of taste and preferences among different national and gender viewer groups. Neither Sky Channel nor Super Channel (the satellite television channels that began in 1987, the attempt to build a transnational audience for satellite television funded by transnational advertisers), have maintained their transnational strategy. Sky Channel, in spite of achieving wider availability to audiences than any other satellite television channel in Europe, has re-orientated itself towards serving a monolingual UK and Ireland audience. Super Channel's future remained uncertain after its rescue from effective bankruptcy by BetaTelevision, however it has continued in service with a mixed programme schedule.

Satellite television has been fraught with risks. Launchers and satellites are unreliable; no audience of a size sufficient to amortise the costs of service can be assured; advertising revenues for international services are insufficient to fund the cost of such services; distribution is dependent largely on cable networks which are not universally available and which do not permit the satellite broadcaster to control either pricing or marketing of services. Nonetheless Europe has not only a large number of established second generation satellite television services, but also a growing number of third generation, direct-to-home services which will, operators hope, transcend the limited access to large homogeneous markets (*ie* the large countries of Europe) that has bedevilled services distributed from low-powered satellites. However as will be clear from the discussion above, it is not only lack of access to audiences that has limited the impact of second generation services. Most important has been the nature of the terrestrial broadcasting services with which satellite television competes. These remain much better funded than the new satellite-delivered entrants to the European television market, are established in the market place and have achieved virtually universal penetration of households. Where their programming is insufficiently attractive to audiences to retain viewers in the face of new competition from satellite television, then readjustment of their programming and scheduling will usually enable them to compete very successfully (as was the case of television in the Netherlands) against the satellite space invaders.

5 European television satellites

The leading trade journal for the new media, *Cable & Satellite Europe* (March, 1989 p. 116), shows 25 satellites transmitting television signals which can be received in the UK. More will be added during the course of 1989 and in years to come. There are also several other satellites that could be added to *Cable and Satellite Europe's* list, but they are relatively unimportant and have had little impact on the core Western European satellite TV market.

Cable and Satellite Europe (March 1989) tabulates 45 different channels of television; these are the channels directed to European viewers and transmitted from 9 of the 25 satellites from which signals can technically be received in the UK. (Like most of the statistics reproduced in this study even these simple numbers are subject to change – a year earlier eight satellites were shown transmitting 39 channels.)

Intelsat	VF2	1.0 °W with 4 channels
Intelsat	VAF11	27.5 °W with 10 channels
Intelsat	VAF12	60.0 °E with 6 channels
Intelsat	VAF13	60.0 °E with 1 channel
Telecom	1F2	5.0 °W with 3 channels
Eutelsat	1F4	13.0 °E with 15 channels
Eutelsat	1F5	10.0 °E with 4 channels
PanAmSat	1F1	45.0 °W with 1 channel
Astra	1A	19.2 °E with 12 channels

Several channels are transmitted from two satellites. Most such duplicated channels were first established on low-powered satellites but now also occupy transponders on Astra in order to provide direct-to-home reception. An exception is Drei Sat (3 Sat) which is transmitted from two low-powered satellites, Eutelsat 1F4 and Intelsat VAF12.

Each of the principal West European television satellites and the channels transmitted from them is described below.

Intelsat VF2 1.0 °W

This satellite transmits channels directed towards Scandinavia: the two television channels of the Swedish public broadcaster Sveriges Radio, the Norwegian service TV Norge and the TV1 service.

Intelsat VAF11 27.5 °W

This satellite currently carries the following channels: BBC1/2 (mix), Childrens' Channel, CNN (Cable News Network), Kindernet, Lifestyle, MTV, Premiere (Film Channel), Satellite Information Services, Screensport, TV3 (Scansat).

The footprint of Intelsat VAF11 is centred on South West England and orientated north east. Its reception area is therefore engineered to include the UK, Northern Europe and Scandinavia. It is the main low powered carrier of anglophone services. It is a backup satellite for telecommunications and all its tv transmissions can be pre-empted if necessary in order to ensure continued telecoms services (see *Cable & Satellite Europe*, 12.1987, pp. 34–37).

Services on Intelsat VAF11

BBC1/2

BBC 1/2 is transmitted in a scrambled PAL format (using the Sat TV 'save' system) to inhibit unauthorized reception. It is principally marketed in Denmark. Its schedule is based on that of BBC1, with imported programming deleted (because rights for such programmes have been accquired by the BBC for UK reception only) and with substitutions from the BBC2 schedule. Transmissions began on 4.6.1987 for eight hours a day to 80,000 homes connected to hybrid cable networks (Cable & Satellite Europe 7.1987 p. 6). Since 1.4.1989 the service has been renamed BBC TV Europe, and has been transmitted from 0700–2400. TVRO owners are able to purchase a decoder and licence for reception for two years for £345.

Screensport/Lifestyle/Kindernet

These services, though listed separately, are carried on the same transponder on the satellite and are transmitted at the same frequency in order to reduce transmission costs. They share W H Smith Television as a principal or sole owner. All are transmitted in PAL unscrambled, and are thus available without charge to TVRO owners, but are intended to be sold as low-pay channels on cable. However access to cable networks in West Berlin (*Cable & Satellite Europe*, 5.1988, p. 8) could be secured only on a no-pay basis, and similar provisions obtain in Ireland. Between June 1985 and December 1986 Lifestyle/Screensport lost £4217 million.

Lifestyle

Lifestyle began transmission in October 1985 and is transmitted from 9 am to 3 pm daily for a female target audience. (The channel is transmitted for 9 hours daily on Astra.) The principal shareholder is W H Smith with 75 per cent ownership (there are minority holdings by the UK ITV companies TVS and Yorkshire TV which were the principal suppliers of Lifestyle's programming).

The intention is to: 'beef up the schedule with romantic dramas or US soaps' (*Cable & Satellite Europe*, 6.1987, p. 24).

Screensport

Screensport began in 1983 and lost £2.5 million in its first year. It is transmitted for 9 hours daily (on Astra and on Intelsat) The channel is owned by W H Smith and ABC (owner of US sports channel ESPN) in the proportions 74.5 per cent, 25.5 per cent. ESPN/ABC increased its holding in Screensport from 3.5 per cent to 25.5 per cent in December 1988 by purchasing shares from W H Smith for £4.4 million, thus valuing Screensport at £20.5 million. It is proposed to extend the audience reach of Screensport by increasing the transmission time and by using extra audio tracks in languages other than English. A French language service began in 1988, known as TV Sport, for which the starting capital of £2 million came from the owners in the following amounts: Screensport 34 per cent; Générale des Eaux 34 per cent; Caisse des depots 10 per cent; Lavizzari 12 per cent.

TV Sport is intended to originate programming that will be of interest to French viewers but which has potential for other European markets; such programming is to account for 5–10 per cent of the Screensport schedule (*Cable & Satellite Europe*, 2.1988, p. 25). There is also a German service: Sport Kanal. Screensport is the most popular of the W H Smith theme channels, and is extensively sponsored. The channel is also transmitted from Astra for direct to home reception (and is to be encrypted in order to secure subscription income from viewers). It will become a 24-hour service with soundtracks in English, French, German and Spanish.

Kindernet

Kindernet began transmission in 1988. Like its sister W H Smith channels it is transmitted unscrambled, but unlike them it is transmitted in Dutch. The service runs from 0700–1000 weekdays and 0700–1100 weekends. Start–up capital is estimated to have been 10 million Guilders and initial ownership was 51 per cent by Télécable Benelux (supplier of childrens' programming to Berlusconi), 29 per cent by W H Smith and 20 per cent by Fuji (producer of computer animated programmes such as *Transformers*). Subsequently ownership changed and W H Smith now holds 49 per cent, Télécable Benelux 26 per cent and Fuji 25 per cent. The channel makes no original programming and acquires much of its schedule from its parent companies. Like Lifestyle and Screensport, Kindernet is marketed as a low-pay channel: an indication of possible revenues is the agreement with the Rijswijk cable network which pays 1 guilder per month per subscriber for Kindernet, Screensport and Lifestyle. Estimated breakeven date for Kindernet is 1991. The managing director of Kindernet, Henk Krop, refers to satellite television as: 'a business where no-one makes any money: it's a matter of surviving and getting a strategic position for the future' (*Cable & Satellite Europe*, 3.1989, p. 30).

W H Smith Channels

W H Smith's strategy for television is to develop: 'one of the major growth areas for the group over the next five years. But we want to develop a European tv business, not a UK TV business' (Francis Baron, Managing Director, W H Smith TV, quoted in *Daily Telegraph*, 18.4.1988, p. 4).

However, for Baron, the European market which his company has entered is fragmented:

'I have serious reservations about whether pan-European services such as Sky Channel or MTV can ever work financially. We have to accept Europe for what it is – a landmass of different languages and cultures – and work within that framework' (Baron, *ibid*).

Accordingly W H Smith's television services are increasingly being offered with sound tracks in different languages: 'We are in the process of setting up a German company to do a German version of Screensport and we intend to do the same in Spain' (Baron, *ibid*).

All W H Smith services are originated from Molinaire-Visions in London; in 1987 W H Smith acquired for £7 million a 51 per cent holding in Molinaire-Visions, with a four-language transmission facility.

Cable News Network (CNN)

CNN, which began its European service in September 1985, exhibits a number of unique characteristics. For a start, it is the only European cable service imported 'cross-strapped' from the USA (*ie*, transmitted from the USA, downlinked to Europe and then uplinked for redistribution in PAL to European audiences). It is also unique in its hybrid consumer/producer service status.

Though CNN's programming is designed for European audiences, its reliance on programming originating from the USA is considerable. CNN's European service began in September 1985. The US signal is redistributed in real time thus limiting the potential appeal of the CNN service (peak hour evening viewing in the USA is the dead of night in Western Europe). European broadcasters pay a flat fee for access to, and use of, the CNN feed (*ie* CNN is a 'producer service' like a news agency). Newspapers also subscribe, paying $500 pm for national and $250 pm for non-national papers. A further source of revenue is the European hotel market: hotels pay approximately $0.105 per room per day for CNN. CNN offers 4 mins of advertising per hour, but CNN's experience is like that of most satellite programme channels: few advertising slots are actually sold.

Satellite Information Services (SIS)

SIS has been transmitted for 3–4 hours daily in scrambled B-MAC from May 1987 to subscribing bookmakers. SIS has a potential customer population of *ca* 9700 UK and 600 Irish betting shops. The channel is owned by the main UK bookmakers; 28 per cent by Ladbroke, 19 per cent by Mecca/Grand Metropolitan, 19 per cent by Coral/Bass, 19 per cent by William Hill/ Sears, 10 per cent by the Race Course Association (RCA), 5 per cent by the Tote. SIS offers live daily coverage of at least two horse racing meetings, and one dog race meeting. It has a 3 tier service with fees ranging from £3650 to £6950 per year per subscribing bookmaker's establishment:

    (1) Basic Service 'Racing Facts'

    (2) Racing Facts + Six (Basic Service + 6 screens of images).

    (3) Racing facts + 12 (Basic Service + 12 screens of images).

Its live coverage is originated by five Outside Broadcast units with access to 59 venues. Capital invested is £40–50 million. When suitable sites for TVROs cannot be secured clients are served by landline from the nearest functioning TVRO. The UK Home Office (the authority that licences off-course gambling) requires, as a condition of licensing betting shops, that all shops in a given area be able to subscribe to Satellite Information Services before service is permitted to commence at any one location. The 1988 annual report of Ladbroke, the major shareholder, states: 'Live television coverage by satellite of horse and greyhound racing, via Satellite Information Services, is currently available in 35 per cent of the division's units in the UK. Virtually all Ladbroke shops should be linked to the service by the end of the year. The screening of live action is expanding the market and significant gains in turnover are planned for 1988' (p. 20).

MTV Europe

MTV (Music Television) was established in the USA in 1981. In 1989 40m US cable homes are claimed to subscribe to the MTV service, which is marketed as delivering a world wide reach of 72 million homes. MTV's promotional material claims that: 'MTV is effective in reaching the youth of the world through the universal language of music'.

MTV Europe began in August 1987 as a 24 hour PAL unscrambled channel owned by:

> 51 per cent Maxwell Communications (Mirror Group Newspapers)
>
> 24 per cent British Telecom
>
> 25 per cent Viacom (the owners of MTV in the USA)
>
> (*Cable & Satellite Europe*, 9.1987, p. 24)

The strategy of MTV is to establish itself as a transnational programme channel building on the relative absence of linguistically specific content in its programme stream, and on the readiness of young viewers to consume exogenous television.

> 'MTV stands to be the first truly international network' (Willard Block, ex-president Viacom, in *TBI*, 1.1, p. 20).
>
> 'Music videos are internationally acceptable. For the bulk of our music programming the words are practically irrelevant' (Tom Freston, MTV Chief Executive Officer, Cited in *TBI*, 1.1, p. 20).

MTV Europe has a programme mix different from the MTV channel screened in the USA, though it draws on Viacom's stock of programmes (animations, concerts *etc*). MTV charges low subscriber fees (35p per cable subscriber per month in the UK), and aims to raise additional funding through advertising and sponsorship. It is rebroadcast terrestrially to viewers in Greece. The channel has paid £1.5 million over 5 years as advance royalties to music video copyright holders. MTV pays royalties to programme suppliers in proportion to advertising revenue secured (*New Media Markets*, 5.8, p. 5 and 5.16, p. 14).

Children's Channel

Children's Channel is transmitted from 0600–1600 Central European Time in PAL unscrambled. The channel shares a transponder and frequency with Premiere (and is encrypted at the head end of Dutch cable networks to secure subscription revenue). Children's Channel began in 1984 and is owned by Starstream, a company which in turn is owned by: British Telecom 22 per cent; Thames TV 22 per cent; D C Thompson 22 per cent; Central TV 22 per cent; Thorn EMI 12 per cent.

The owners' investment from 1984 to 1987 amounted to £6.5 million. In February 1989 it was reported (*NMM*, 1.2.1989, p. 6) that the US cable company, United Cable, was to acquire 20 per cent of the channel for £3.5–4.0m. 30 per cent of programming is produced by the channel in studios leased from TV1. The average cost of such programming is estimated to be £5000 per hour. Some programmes produced by Children's Channel have been sold to other broadcasters, notably *Bob's Your Uncle* to US schools. The channel is funded by a mixture of revenue sources, including advertising and sponsorship (*eg* the programme *Glow Boy* by Ready Brek) and by subscription. A representative subscription rate was reported to be that charged in Sweden of SKr 25 per month per subscriber (£2.50) (*Cable & Satellite Europe*, 6.1987, p. 18).

Premiere

Premiere is transmitted from 1500–0300. It is a film channel transmitted in an encrypted form of PAL using the (reputedly easy to hack) SAVE system. The service began in September 1984.

An authorized Premiere decoder costs £124 + VAT and a further annual subscription of £75. The average subscription to Premiere via cable is *ca* £7 per month. In spite of revenues of approx £6.5 million pa Premiere is still in the red, though with a reported positive cash flow in 1988. Transponder costs for the channel are reported to be £1.5 million pa. Where possible the channel has acquired films on a royalty-per-subscriber basis with rights for 10 screenings in 12 months. Premiere began as a film channel titled TEG (The Entertainment Group) with majority shareholdings by Thorn EMI (41.2 per cent) and Goldcrest (9.8 per cent) and minority holdings by Fox, Columbia, HBO and Showtime. The original shareholders were owners of film properties wishing to explore a new medium of distribution. In April 1986 Maxwell Communications acquired the holdings of Thorn EMI and Goldcrest, and became the majority shareholder in TEG with 51 per cent. Maxwell Communications merged TEG with the rival satellite film channel Mirrorvision (which had earlier subsumed another competing film channel The Entertainment Network – TEN – The Movie Channel) to create a single UK satellite-delivered subscription film channel.

From 1987 ownership of Premiere was: British Telecom 30 per cent; Maxwell Communications 30 per cent; Fox (owned by Rupert Murdoch's News Corporation) 10 per cent; Columbia Pictures 10 per cent; Home Box Office 10 per cent; Showtime 10 per cent.

Premiere is scheduled to close down in late 1989 when its rights to film properties expire. Its owners declined to pay the prices paid by BSB and Sky Movies for product to enable the company to continue in existence and compete against these two new film channels orientated to UK viewers and using DBS.

TV3 ScanSat

TV3 Scansat is transmitted in B-MAC for 4–6 hours daily. TV3 began service in December 1987 as a general TV programme channel for viewers in Scandinavia designed to offer: 'The best of American television with a touch of ITV's philosophy and the quality of the BBC' (Jan Stein Mann, TV3 president, *Cable & Satellite Europe*, 2.1988, p. 34).

TV3 claims to produce 20 per cent of its programming and has programme supply agreements with Disney, Columbia, MCA, New World Pictures, Fox, Pathé, Bavaria, Thames Television and Central Television. For TV3's target audience in Scandinavia, Swedish is both the language most readily understood by Scandinavian non-Swedish speakers and the language of the largest linguistic group. It is ScanSat's preferred language of programming. All ScanSat programmes not of Scandinavian origin are dubbed into a Scandinavian language – usually Swedish. Programmes of Scandinavian origin will be subtitled in one of the other two main Scandinavian languages. News programmes will be spoken by Swedish *and* Danish presenters. TV3 is owned 96 per cent by Kinnevik (Sweden) and 4 per cent by Nora (Norway). The channel is funded by the sale of advertising, and currently enjoys a monopoly of television advertising in Sweden. In Denmark and Norway terrestrial broadcasters compete with TV3 for advertising revenue, but in both these countries television advertising is a recent phenomenon, and there are therefore no strongly established relationships between advertisers and terrestrial broadcasters.

The channel is uplinked from the UK, and its main offices are in London, 'the television capital', to facilitate acquisition of programme rights etc. TV3 is also transmitted from Astra (in which Kinnevik has a financial interest). A second Scansat channel is projected. TV3 has access to 90 per cent of Scandinavian cable homes (*ca* 20 per cent of Norwegian, 11 per cent of Swedish, and 35 per cent of Danish homes are cabled) and achieves a daily reach of approximately 30 per cent of viewers with access to the channel. In 1988 TV3 generated £5.4 million income, against outgoings of £14m, (of which £12.6 million was spent on programming). Break-even is anticipated in 1992 (*Cable & Satellite Europe*, 3.1989, p. 8).

The channels described above are currently (March 1989) transmitted from Intelsat VAF 11. Other channels have occupied transponder capacity on this satellite before ceasing service. The following two services are now no longer transmitted, but exemplify interesting *failures* to utilize the transnational capability of satellite television. 'Transnational' both in the sense (exemplified by ScanSat and Canal 10) of transmitting signals to an audience in one state from the territory of another, in this case from the UK to the Scandinavian countries and to Spain, and also in the sense of reaching an audience dispersed over the territories of several states.

Canal 10 (Film Success)

Canal 10 was a Spanish language pay channel uplinked unscrambled from the UK. It began in January 1988 and ceased transmission in August of that year. It was developed by Spanish interests unsuccessful in securing terrestrial TV licences from the government of Spain. Some programming was in English and it was hoped the 'reach' of the channel would extend to the British expatriate community in Spain. The transponder rented from British Telecom International and used to transmit Canal 10 had the capability of transmitting two sound tracks: one Spanish, one English. The channel offered 24 hour entertainment: films, soaps, documentaries, pop music and sport. Canal 10 had programme supply agreements with US film majors and with the UK broadcasters Granada, Thames, TVS, Tyne Tees. The channel estimated its breakeven level at 200,000 subscribers and had a target of 800,000 subscribers. However, it was reputed never to have had more than 654 subscribers.

The ownership of Canal 10 was: Talar 46 per cent (Spanish film Distributor); Noara 15 per cent (Spanish holding company); Caha de Aorras de Vittoria 12 per cent (Spanish bank); Lahenood 7 per cent (Finance House); H Capital 5 per cent (Spanish finance house); CLT (Luxembourg) 5 per cent; Canal Plus 5 per cent.

The holding company, Film Success International, was registered in Panama. The shareholders accused the channel's founder of fraud. Canal 10 ceased transmission when the UK facilities house Molinaire suspended its services after its bills were unpaid.

The *EBU Review* (March 1988, vol. XXXIX, no. 2, p. 33) reported an ownership change in Canal 10 to 10 per cent ownership by Maxwell Communications, an increase in its stake to 10 per cent by Canal Plus and a similar increase to 10 per cent by CLT.

Anglovision

Anglovision began on an experimental basis in June 1987 with full service planned for September. Full service was never established. The experiment, of a programme service for anglophone travellers, downlinked to several Paris hotels (6–7000 rooms), included 6 hours of news and current affairs from NBC and PBS (*Cable & Satellite Europe*, 7.1987).

Ownership was: NBC News International 30 per cent; Irish Independent Newspapers, Herald Anglovision 30 per cent; American Express Venture Associates 30 per cent; Andy Mulligan (Anglovision President) 10 per cent. Service was free to hotels and was to be financed by advertising.

The Discovery Channel

An additional new service, The Discovery Channel (a Europeanized version of a well established and successful US channel scheduling 'fascinating fact' adventure and exotic documentaries) is scheduled to open on Intelsat VAF11 in 1989.

Intelsat VAF12 60.0 °E

The principal satellite for German language services, its footprint being centred on Nuremberg. The satellite transmits the following channels: AFRTS Germany; West 3; BR 3; 3 Sat; Eins Plus, Pro 7.

An important mode of access to German language satellite channels (whether delivered by Intelsat or Eutelsat satellites) is via terrestrial re-broadcasting. Consequently West German viewers enjoy access to satellite television (via terrestrial re-broadcasting) beyond the extent suggested by the number (more impressive than in any other large European state) of German cable connections.

AFRTS Germany

The English language American Forces Radio and Television Service-AFRTS is transmitted 24 hours daily in B-MAC 525 lines. The service was established in November 1987 as a means of distributing to US forces stationed in Europe a pot pourri of broadcast television from the United States (a compilation derived from the schedules of the three principal networks: ABC, CBS and NBC).

Pro 7 (Formerly Eureka)

Pro 7 is transmitted for 12 hours daily in PAL unscrambled. The channel is funded by advertising, (estimated revenue DM 800,000 per month in early 1989). In 1988 its ownership and programming orientation changed. From a schedule based on re-transmission of World Net (a service of the United States Information Agency) and religious programming, Pro 7 changed to general entertainment orientated to affluent 'yuppies'. 49 per cent of the ownership of the channel is vested in Beta Taurus, the West German film distributor; other important interests are held by Ackerman supermarkets.

West 3 (formerly WDR 3)

A German language service transmitted for 10 hours daily in PAL unscrambled. West 3 uses the satellite to redistribute the third channel of West Deutscher Rundfunk (the largest member of West Germany's ARD public network) as a West German 'superstation'.

BR3

A German language service transmitted in PAL, unscrambled, for 10 hours daily. Bayer ischer Rundfunk's third channel 'superstation'.

3 Sat (Drei Sat)

3 Sat (Drei Sat) is a German language channel composed of programming from West German (ZDF), Austrian (ORF) and Swiss (SRG) public broadcasters. The channel is transmitted from 17.25 to closedown at about 2400. Programming is largely high culture and political documentary and comment. The Channel is also transmitted on Eutelsat 1F4.

Eins Plus

Transmitted 4–5 hours daily in PAL unscrambled and programmed by the first West German public broadcasting network ARD. There are some differences between the programming of the terrestial ARD service and Eins Plus but also many exact parallels (*eg* the same news is transmitted at 8 pm on both channels). Essentially Eins Plus is a satellite-delivered echo of a terrestrial channel.

Eins Plus, Drei Sat, BR3 and West 3 owe their existence to the desire of West German broadcasters to 'balance' commercial satellite television channels with public channels. Therefore RTL Plus, Tele 5, TeleClub and Sat Eins (transmitted from Eutelsat 1F4) are 'balanced' by satellite re-transmission of programme schedules constructed by the public broadcasters of West Germany. Not surprisingly, the satellite channels significantly replicate terrestrial broadcasts. The public broadcast network, ARD, programmes one satellite channel (Eins Plus). The ZDF, the second West German public television network (in conjunction with neighbour country German language public broadcasters) transmits another (3 Sat). The nominally regional 'third' channels of the public broadcasters of the *Länder* of Bavaria and North Rhine Westphalia, BR3 and West 3, are elevated to the status of 'superstations' receivable via satellite over the whole of West Germany (and beyond). Thus commercial satellite television is 'balanced' by an equivalent number of public channels. There is also a significant balance struck between BR3 and West 3, in that the former channel emenates from a public broadcaster associated with political control by West German conservatives, while the latter has links with Social Democrats.

Intelsat VAF13

Although a different satellite to Intelsat VAF12, VAF13 is 'parked' at the same orbital location and signals from it are receivable using the same antenna as that used for Intelsat VAF12. The Tele 5 service transmitted from VAF13 was formerly transmitted from VAF12; the change of satellite will have made no difference to reception.

Tele 5 (formerly known as KMP and Music Box)

A music channel transmitted 24 hours daily in PAL unscrambled by KMP (Kabel Media Program Gesellschaft) owned: 45 per cent Berlusconi; 45 per cent Tele München (Kloiber); 10 per cent Wolfgang Fischer (general manager KMP); formerly owned Fischer 5 per cent, Kloiber 25 per cent, Bauer 25 per cent, Burda 25 per cent, Communication & Entertainment Ltd (Australia) 20 per cent (*Cable & Satellite Europe*, 5.1987, p. 20).

Tele 5 programmes music, features, sport and entertainment interspersed with 5 minute newscasts (in collaboration with CNN) transmitted on the hour. The budget for 1988 was $35 million. Tele 5 has regular access to 4 million cable homes and occasional access to a further 3.4 million via terrestrial re-broadcasting (*NMM*, 1.2.1989, p. 12) The channel is reported to experience difficulty in raising advertising revenue, especially after its shift to a general programming format from its former music format (*Variety*, 10.2.1988, pp. 91–114).

Telecom 1F2 5

Telecom carries the majority of French language satellite TV channels. M6 and La Cinq are principally terrestrial services for which satellite rather than terrestrial microwave relay is used to feed terrestrial re-broadcasting transmitters.

Canal J

Canal des Jeunes (Children's Channel) began transmission in January 1986 as a French-language children's service, transmitted for 10 hours daily in PAL unscrambled. Canal J is funded by a subscription of 5FF per subscriber per month, and by advertising of up to 6 minutes per hour (but as with most satellite television channels advertising has proven difficult to secure; only 1½–2 min per hour have been sold to *inter alia* Kellogs and Mattel). Canal J is 100 per cent owned by Hachette. It has a novel format, in that it shows only children on screen. It transmits 10 news programmes per week. 15 per cent of its programming comes from Hachette: 35 per cent of its programmes are of francophone origin (the channel has rights to Tintin and Babar television programmes), and its non-Francophone programmes are dubbed into French. Its programme budget is *ca* £1.65

million pa. Canal J has agreements to be carried as the sole children's channel by the main French cable networks (Générale des Eaux, Lyonnais des Eaux, Caisse des Dépôts et Consignation). Canal J is marketed only in metropolitan France and is excluded from cable networks in Wallonie and Romandie as the transponder used for the channel is rented on a domestic rate (*Cable & Satellite Europe*, 4.1987, p. 62).

La Cinq

La Cinq is transmitted for 24 hours daily in Secam unscrambled. It is a terrestrial service which uses satellite as a means of linking cable networks and, more important, terrestrial transmitters. La Cinq has enjoyed a rather precarious existence with low ratings and disagreements over programming policy among its owners. Its low ratings in 1987 (a 7.4 per cent share) occasioned criticism by one owner, Berlusconi, that the channel screened too much news emanating from another owner, Hersant, and insufficient entertainment. La Cinq has experienced the contradiction that has beset several satellite channels: scheduling (soft) pornographic films raises ratings but deters advertisers who do not wish their products to be associated with controversial programming.

Ownership 25 per cent Hersant (French publishing conglomerate): 25 per cent Berlusconi (Rete Italia); 10 per cent Chargeurs (Seydoux); 15 per cent Pargeco (Mutualite Agricole); 9 per cent Societe Centrale d'Investissement; 5 per cent Les Echos (French newspaper company with ownership by UK Pearson conglomerate); 5 per cent Credit Lyonnais; 5 per cent TéléMetropole (Québecois commercial broadcaster).

M6

French music channel transmitted for 10 hours daily in Secam unscrambled. A terrestrial service re-distributed via satellite achieving approximately a 6 per cent share of the French television audience. It is best known for its *Sexy Clips* programme. Minority shareholder CLT 25 per cent.

Eutelsat 1F4 13 °W

Eutelsat 1F4 is the 'hottest' of low powered European TV satellites, as it carries the most important channels for the West German market, RTL Plus and Sat Eins; the two Anglophone channels which have achieved the highest penetration of European households, Sky Channel and Super Channel; and the most successful European film channel, FilmNet.

RTL Plus

RTL Plus began as a terrestrial broadcast service in German transmitted from Luxembourg by CLT (Compagnie Luxembourgeois de Télévision). It extended its coverage area across West Germany by opting for satellite distribution. Viewers access RTL Plus via both cable and terrestrial re-broadcasting. RTL Plus is transmitted for 8 hours daily in PAL unscrambled. The company domicile is now in West Germany, reflecting both majority ownership by West German interests and a desire to secure favourable allocation of terrestrial re-broadcast frequencies from the Bundespost.

Ownership is: 46 per cent CLT; 50 per cent West German press interests (Bertlesmann 39 per cent, Frankfurter Allgemeiner Zeitung 1 per cent, West Deutscher Allgemeiner Zeitung 10 per cent); 4 per cent by the Deutsche Bank (possibly as a proxy for the Suddeutscher Verlag).

In 1987 the channel reached 2.7 million homes, principally in West Germany (including 400,000 receiving terrestrial broadcasts from Luxembourg), and in 1988 8.8m homes following acquisition of terrestrial rebroadcast frequencies (4.6 million homes via re-roadcasting) and extension of West German cable networks (4.2 million homes via cable).

RTL Plus carries a maximum of 12 minutes advertising per hour orientated to the German market and has a reputation for addressing a youthful and downmarket audience; it achieves a 4.1 per cent share in the whole West German market and 12.5 per cent in cable homes.

RTL Plus is reported to estimate an income of DM270 million (*Cable & Satellite Europe*, 3.1989, p. 47) and DM200 million (*NMM*, 1.2.1989, p. 12) in advertising revenue in 1989. RTL anticipate breakeven in 1992, having increased its advertising rates by 79 per cent between 1988 and 1989.

Advertising income has risen as follows:
    1988 DM110 million
    1987 DM45–50 million
    1986 DM27.5 million
    1985 DM15.3 million

RTL Plus pays DM2 million per annum to the Bundespost for transponder rental and DM3 per household served. 45 per cent of RTL Plus programmes are produced in-house (including a growing number of local programmes for particular markets served) and its news services (doubtless strengthened by the parent company CLT's membership of the EBU, with consequent access to Eurovision news exchanges) have, in at least one West German market (Saarland) attracted a majority of the audience for TV news. The percentage share is: RTL Plus News 42.6 per cent, Tagesschau (ARD) 26.4 per cent, Heute (ZDF) 26.9 per cent. The lynch pins of the RTL Plus schedule are feature films (from the US, West Germany and Italy in the main) and US TV series (*Cable & Satellite Europe*, 5.1987; *Cable & Satellite Europe*, 12.1987, pp. 32–33).

Sat Eins

Sat Eins began as a German language service on the experimental Ludwigshafen cable system in 1985 and then began national distribution through West Germany via satellite. It is owned by West German press interests: PKS 40 per cent; Springer 15 per cent (itself 25 per cent owned by Leo Kirch, an influential interest behind Pro 7); Holtzbrinck 15 per cent; Neue Median Gesellschaft Ulm 1 per cent.; Otto Maier 1 per cent. The remaining portion of the company is owned by a consortium of 138 West German newspapers, Aktuell Press Fernsehen (APF) with 15 per cent and an unspecified holding company with 13 per cent.

Sat Eins transmits in unscrambled PAL standard for 18 hours daily. Sat Eins is headquartered in Mainz, this siting being a condition of gaining access to cable networks under the jurisdiction of Land Rhine Pfalz. It produces programmes in Ludwigshafen and West Berlin. To secure access to terrestrial frequencies in West Berlin Sat Eins undertook to assume 51 per cent ownership of a West Berlin production home and to locate its programme production (except news) and dubbing in Berlin. In 1986 Sat Eins budget was DM250 million; in 1987 DM150 million (of which 100 million was spent on programming, 30 million on News, 20 million on Administration). Its transponder costs DM2 million + DM3 per household per annum. (This came to a total of *ca* DM8 per household in 1987.) It pays the Deutsche Bundespost about DM6200 for annual rental of each 2000 watt transmitter used for terrestrial rebroadcasting. Sat Eins has access to 8 million West German homes; 4.2 million via cable, 3.8 million via terrestrial rebroadcasting. It achieved a 21.9 per cent share of viewing in cable homes in 1988 and a 5.6 per cent share of the whole West German television audience in 1988 (*NMM*, 1.2.1989, p. 12).

Estimated revenue has been in 1989 DM250 million, in 1988 DM118 million, in 1987 DM35 million, in 1986 DM16 million, and in 1985 DM5.9 million.

In 1987 Sat Eins charged DM 1200 for national coverage of a 7 second advertising slot. This compare with DM1116 charged for a comparable slot on the NDR region of the ARD network, delivering audiences in the localities of Hamburg, Schleswig-Holstein and Nieder Sachsen. The NDR slot, though cheaper, will usually deliver a higher audience share in its universe than will the Sat Eins advertisment in its (*Cable & Satellite Europe*, 5.1987). Even so, in 1989 Sat Eins increased its prime time rates by 88 per cent over 1988 levels. It anticipates breakeven in 1990. The Sat Eins programme schedule is, like that of its competitor RTL Plus, built around feature films (from USA, UK, West Germany) and US tele-series.

Teleclub

Teleclub began as a German language film channel in Switzerland in 1984. It is transmitted unscrambled in PAL (and scrambled at cable head ends) for 6 hours daily. Consumption is centred on Switzerland with, in addition, about 1000 German subscribers on a single cable network in Hanover, West Germany. Penetration in Zürich is approx 9 per cent of the potential audience, in Hanover *ca* 5 per cent. Teleclub is owned 40 per cent by Beta Taurus (the leading West German film distributor) and 60 per cent by Rediffusion AG. Beta Taurus has German rights to feature films from Columbia, Warner, Fox and Paramount. Teleclub shows few Swiss films since most are co-financed by SRG (Swiss Broadcasting) and are therefore unavailable to competing broadcasters. The West Germany Teleclub is owned by Beta Taurus, Bertlesmann and Springer and operated by Beta Taurus. Swiss charges are SFr28 per month, of which SFr8 goes to the cable operator whose network distributes the channel, and SFr20 accrue to Teleclub. In West Germany Teleclub costs DM29 per month (for comparison, the basic cable charge in West Germany is *ca* DM9 per month). Subscribers pay an additional decoder deposit (Switzerland SFr100). Breakeven is estimated at 80–90,000 subscribers (*Cable & Satellite Europe*, 12.1987, pp. 16–17).

TV5

TV5, a French language channel, began operating in 1984, and is now transmitted daily from 1600–2400 in PAL unscrambled. It draws on television programmes from the main French network channels (TF1, A2, FR3), and from Belgium (RTBF), Canada (CTQC Consortium de télévision du Québec-Canada, *ie* programmes from Radio Canada and Radio Québec) and Switzerland (SSR). TV5 'aims to strengthen the position of the French-speaking world' and 'to give people a francophone view of the world'. It reaches 5.5 million cable homes in 20 countries and plans to extend service to Canada (*EBU Review*, 1.1988, pp. 30–31).

Worldnet

Worldnet, ran by the United States Information Agency, ceased transmission of its news and information programming after the United States Congress withdrew funding. If a Congressional appropriation is renewed the service will resume. Worldnet shared a transponder with TV5 and was transmitted 0700–0900 and 1400–1600 Central European Time for 5 days a week. It had an annual budget of $19 million, $4 million of which went to its European service Euronet. It offered a mixed schedule of news, sport, science, culture and English language programmes and carried no advertisements. The objective of Worldnet was defined as: 'To provide a service that represents how people live and work in America; and to have a dialogue with Europe once a day.' A. L Snyder, Director (*Cable & Satellite Europe*, 7.1987, p. 29).

FilmNet

A 24 hour PAL scrambled network, FilmNet is the biggest European film channel, reaching 213,000 homes. Of homes having access to it, the channel has achieved penetration of *ca* 3.75 per cent in the Netherlands, 3.25 per cent in Belgium, and 7.5 per cent in Sweden. The FilmNet service is subtitled, using teletext, in Dutch, Danish, Swedish, Finnish and Norwegian. Operators were reputedly 'surprised to discover how important subtitling was'.

FilmNet is sold in three tiers:

Access between

    0600–1400 Skr 90 per month (*ca* £9)

    0600–2200 Skr 110 per month

    0600–0600 Skr 125 per month

    Decoder rental 35 Skr per month

Two erotic films are shown each week. They are screened late at night and can only be accessed on the premium tier. More than 50 per cent of FilmNet subscribers choose 24 hour access. One new title is shown every day. Payments to distributors for film properties is variously on a flat fee or per subscriber basis. Per subscriber fees range from £0.005 (*ie ca* £1000) to £0.15 (*ie ca* £3000) per film. FilmNet pays unspecified carriage fees to cable networks. FilmNet is owned 80 per cent by Esselte (a Swedish video distributor) and 20 per cent by Rob Houver (UIP) a Dutch film producer. An earlier partner, the Dutch publisher VNU, withdrew in 1986 after FilmNet sustained £3.5m losses in 1985 (*Cable & Satellite Europe*, 12.1987, p. 18–21).

3 Sat

See entry under Intelsat VAF12.

A number of services listed as channels are in fact transmitted both as independent programme streams, that is as distinct channels, and as successive components of an integrated 24 hour programme stream. Thus the programme streams listed below may be transmitted successively from transponder 6 on Eutelsat 1F4 to create the illusion of a single channel, and also as distinct programme streams, from Astra and Eutelsat.

From Transponder 6 Eutelsat 1F4	
Sky Channel	0600–1300: 1400-1700
Arts Channel	0030–0330
EBC (European Business Channel)	(2x1/2 hrs in Sky Channel)
Landscape Channel	1300–1400; 0330–0600
Eurosport	1700–0030

Sky Channel

Sky Channel was the first European satellite TV channel and began transmission from the UK on Orbital Test Satellite 2 (OTS2) in April 1982. The company was established as Satellite Television plc by Brian Haynes (ex Thames TV) backed by Guinness Mahon, Barclays, Ladbroke, D. C. Thompson and others. In 1983 News International purchased 65 per cent of Satellite Television, and from 1988 has owned 82 per cent of the Channel. In 1987 Sky raised £22.63 million in a rights issue (in addition to previous rights issues of £5.29 million) and further rights issues are anticipated as a consequence of the heavy expenditures entailed following Sky Channel's movement to direct-to-home broadcasting from the Astra satellite. The channel has incurred growing losses throughout its history:

In the 15 months to June 1984, a pre tax loss of £5,769,621
In the 12 months to June 1985, a pre tax loss of £8,638,700
In the 12 months to June 1986, a pre tax loss of £5,685,081
In the 12 months to June 1987, a pre tax loss of £10,198,801
In the 12 months to June 1988, a pre tax loss of £8,452,544

Sky had an accumulated pre-tax deficit to 30 June 1988 of £38,744,747. Its growing losses are attributed to: 'increased competition in a developing marketplace and continued difficulties in obtaining entry and exercising Sky's full market potential, in some key European territories' (*Cable & Satellite Europe*, 11.1987, p. 5).

Sky has also incurred costs by decreasing the proportion of purchased programming (from 59 per cent in 1985/6 to 56 per cent in 1986/7) in favour of increasing its own production, and by 'buying' its entry to Dutch and Belgian cable systems. Sky has paid 'carriage fees' in kind by purchasing an agreed value of programming (£240,000 to gain access to Walloon cable nets) or establishing production units (Nederlands Instituut voor Lokale Omroep NILO, approx £800,000) (*Cable & Satellite Europe*, 11.1987, p. 30). Sky's 'Pop Formule' music show (07.35–08.35 Monday mornings) is made with the Dutch broadcasting society TROS, and its weekday morning *D J Kat Show* is produced by John de Mol Productions in Loosdrecht. Sky has also concluded local production agreements in Paris and West Berlin.

Sky's advertising revenues increased from £7.8 million in 1985/6, to £9.4 million in 1986/7, and to £12.1 million in 1987/8. As well as conventional spot advertising (limited to the IBA norm of 7 mins per clock hour) Sky also screens sponsored programmes (*eg* by American Express, Coca Cola, TDK, Canon, Gillette, Ford, Audi) including the Uniroyal weather report, golf sponsored by the Spanish tourist board, and a morning home shopping show linked to *Knobs and Knockers* and *Discounter*).

Sky has been transmitted for differing periods ranging from 18 hrs to the current 12 hrs (on Eutelsat) a day. The channel has been sometimes transmitted in scrambled form in order to satisfy its programme distributors and performing rights bodies that its programmes cannot be pirated and redistributed illegally. More commonly, it has been transmitted unscrambled because of 'technical difficulties' or 'decoder shortages'. At the time of writing (March 1989), the channel is transmitted unscrambled.

Sky is carried on nearly all European cable networks (except in Flanders which has a high cable penetration and is therefore a significant market foregone by Sky Channel) and is close to achieving the maximum possible availability for a satellite TV channel. Like other satellite channels Sky has experienced difficulty in reconciling the desires of advertisers and of audiences. Its highly rated show *Wrestlemania* scored a 3.4 per cent viewing share but did not attract advertising because it was perceived by advertisers as being too downmarket for their purposes.

Sky Channel offers what it describes as 'alternative' television: 'In the UK and other highly developed TV environments (USA, Australia, Japan) alternative would mean minority public service broadcasting. Where Sky is most popular this kind of broadcasting is the mainstream and a low brow general entertainment channel becomes the preferred choice' (*Cable & Satellite Europe*, 9.1987, pp. 14/15).

In 1987 Sky claimed that 56 per cent of its programming was of European origin (of which nearly half was produced by Sky itself), 32 per cent from the USA and 12 per cent from the Commonwealth. It further claimed to have 'developed programming and scheduling into a detailed science' (*Cable & Satellite Europe*, 9.1987, p. 14).

Before 1989 and the inception of its service on the Astra satellite, designed for direct reception in UK and Irish homes, Sky Channel attempted to construct a transnational European audience. To realize its 'pan-European programming philosophy', Sky produced its music/video shows in London, the Netherlands and Belgium. It also took a feed from the West German KMP music channel, and collaborated with the Canadian music video channel Much Music/Musique Plus. However this attempt to build a pan-European service was unsuccessful, and caused Sky to retreat. Its new direct-to-home format, transmitted from Astra, has the UK and Ireland as target markets. In its new focus on a monoglot Anglophone audience Sky Channel's annual programming budget is estimated at £30 million per annum (*NMM*, 7.1.1989, p. 3).

Arts Channel

The Arts Channel began in 1985, and was first transmitted from Intelsat VAF11. The Channel has been close to closure at several points (its principal shareholder W H Smith divested itself of its 39.4 per cent holding in 1988), but is now claimed to be near breakeven after investment from United Cable Television. The channel was transmitted from 1700–1000 Central European Time when carried on Intelsat VAF11 and when sharing the Sky Channel transponder (from early 1988) from 0030–0330. It carries no advertisements and defrays programming costs by carriage fees from cable operators, pre-sales to broadcasters, and sponsorship (including sponsorship from British Gas, Irish Tourist Board, *Country Life*). The channel's schedule is made up of documentaries and visual arts (36.5 per cent), classical music (25 per cent), opera (15 per cent), drama (10 per cent), dance (7.5 per cent) and jazz (6 per cent). Average programme cost per hour is estimated at £2000. The channel is widely available, since it shares a transponder with Sky Channel, to which it lends high cultural legitimacy.

EBC (European Business Channel)

EBC was established in late 1988 as a dual language 'thematic' channel with a target audience of Anglophone and German speaking businessmen. EBC offers two services: as a distinct channel it transmits two hours daily in German (sharing the Teleclub transponder), while in English, as a component of the Sky Channel programme stream, it provides two half hour daily slots. A German language service similar to that delivered to Sky Channel and embedded in the programming of either RTL Plus or Sat Eins is anticipated (Sat Eins was a backer of EBC when the channel was first proposed but withdrew from the venture before its launch). Both the ownership of EBC (a consortium of Swiss interests) and the financial arrangements with its host channels, are unclear. It claims a £6m annual production budget (*Cable & Satellite Europe*, 11.1988, pp. 60/61).

Eurosport

Eurosport is run as a joint venture between Sky Television and public broadcaster members of the European Broadcasting Union (EBU).

The members are: Sky Channel, RUV (Iceland), BBC (UK), RAI (Italy), SRG (Switzerland), RTBF and BRT (Belgium), SVT (Sweden), YLE (Finland), NRK (Norway), Danmarks Radio (Denmark), ERT (Greece), ORF (Austria), JRT (Yugoslavia) and RTE (Ireland).

The Eurosport consortium negotiates with sport promoters and associations for coverage rights, and records and transmits coverage of events. Members of Eurosport have access to coverage of events to which Eurosport has secured rights 48 hours before access is given to non-members. The early schedules of Eurosport (February 1989) have included substantial amounts of programme material from non-Eurosport members – from ABC in the USA, from ITV and from public broadcasters in West Germany, Italy and France. Broadcaster members of Eurosport may transmit coverage of events in a different format and at different times to that transmitted in the Eurosport satellite television programme stream. Eurosport has a six hour schedule which is repeated twice on Astra (giving a daily transmission time of 18 hours) and transmitted once on Eutelsat. The estimated annual cost of Eurosport is *ca* £30m of which half will be incurred by Sky Television (*NMM*, 7.1.1989, p. 3). W H Smith Television (Screensport) has secured an opinion from the European Commission that, prima facie, Eurosport breaches EEC competition law. It is likely, therefore, that Eurosport's command of access to events will be successfully challenged in a European court and the channel will cease to exist in its present form. Clearly the commercial future of Screensport, marketed as a subscription channel, is severely compromised by the continuing existence of Eurosport as an advertising-funded 'free' channel.

Landscape Channel

The Landscape Channel juxtaposes video images of beautiful and striking landscapes with a sound track of instrumental music. It is funded by revenue from direct mail sales of recordings of the music played on the channel. A subordinate revenue flow is the fee of 5p per subscriber per month demanded of cable networks that re-distribute the channel (offset by a 5 per cent commission to cable operators on music sales). The channel began in October 1988 and is transmitted for 2–3 hours nightly in the closed–down period of Sky Channel transmissions.

Super Channel

Super Channel began transmission in Feb 1987. It is on air up to 20 hrs daily from 0700-0300 and transmitted in English unscrambled in PAL. Super Channel began as Music Box owned by Granada, Virgin and Yorkshire TV. It became a general programme channel following the conclusion that: 'The universe of homes is not big enough to support thematic channels' (R. Hooper, Managing Director, Super Channel, *Cable & Satellite Europe*, 9.1987, p. 24).

In 1986 ownership passed to the UK ITV companies (excepting Thames Television) and to the Virgin company. The programme schedule changed to a mixed schedule (including pop music and videos) designed to bring to European audiences the *Best of British* television. The channel claimed to have orientated its programming to its target audience: 'Super Channel takes into account that most viewers are not native English speakers. Presenters speak clearly, comedies and documentaries are selected for their visual content while music and sports programmes have a universal appeal'. (Super Channel Information Pack, 1988)

But the research carried out by PETAR on Super Channel's audience suggested that insufficient attention had been given to the linguistic competences of the target transnational audience. *The Best of British* strategy proved no more successful than had the Music Box initiative and, in an attempt to attract Western European viewers, the channel scheduled occasional subtitled programmes sourced from Western Europe (for example West Germany's *Tatort* and the Netherlands *Say Ah*). These metamorphoses in Super Channel's programming and scheduling meant that the channel never established a consistent and coherent identity, and consequently viewers were unable to form confident expectations about the channel's programme offer. The succession of format and content changes reflected the difficulty of establishing a financially viable corporate strategy for satellite television in Western Europe with either a thematic or general interest channel.

In early 1988 Virgin became the largest shareholder in Super Channel with 35 per cent (the ITV companies declining to take up their share in rights issues designed to recapitalize the failing company). Losses in 1988 were estimated to be £8 million (*Financial Times*, 25.3.1988, p. 8). Accumulated losses in the channel were estimated at £46 million.

Later in 1988 Super Channel was effectively bankrupt: its ITV owners ceded their shareholdings to Virgin whose holding increased to 45 per cent, and to an Italian interest, Beta Television, (which became the majority shareholder with 55 per cent), having purchased its share of Super Channel for £1 and assumed debts of approx £7–8 million, Beta Television settled with the creditors of Super Channel for 25p in the £1. Super Channel's identity after the Beta-Television takeover remains unclear. Its schedule is described in *Satellite Times* (a programme guide) as 'General entertainment'. Future plans include an emphasis on music directed towards a 'young adult' audience. Present programmes are mainly old favourites such as *Benny Hill*, *Dr Who*, *Dempsey and Makepeace* and *The Professionals*. This schedule appears to retain all the lack of focus of its previous incarnations. Super Channel's programming has come principally from the ITV companies and the BBC and included a European News Service from ITN and European weather from the UK Met Office and the BBC.

Super Channel's *Best of British* profile was more honoured in the breach than the observance, due to the absence of an agreement between the channel and the UK talent unions on a formula to compensate those who had worked on the archive programmes Super Channel wished to re-transmit. Super Channel drew on programmes owned by its shareholders and which they had written off in their first year of production; but substantial costs attached to re-transmission of such programmes due to the right to residual payments established by the UK unions on behalf of their members. In Super Channel's first six months, during which the *Best of British* strategy was followed, the channel incurred costs of £385,000 to Equity in residual payments to actors (*NMM*, 5.25.19 p. 9). Subsequently Super Channel's schedule included many US and Australian programmes. Like Sky Channel's experience with *Wrestlemania*, the American and Australian programmes were reasonably successful in attracting viewers but not in attracting advertisers, because the ABC1 audience desired by advertisers tended not to watch such programmes.

Super Channel revenues derived principally from advertising but in volumes insufficient to cover costs. It enjoyed some success in finding sponsors for programmes. Goodyear sponsored the daily weather report and other sponsors included British Airways, Pepsi Cola, Philips, Gillete and Toyota. The channel has maintained its strategy of building a transnational audience, and from April 1989 began transmitting most (*ca* 65 per cent) of its programming with alternative sound tracks in English, Dutch and German.

Pace (Programme for advanced continuing education)

An experimental programme in international technical education. It was previously located on Eutelsat 1F2, (a satellite used for a variety of television 'producer services' such as Eurovision programme exchanges and Visnews and WTN news feeds).

Galavision

A 24-hour Spanish language service originated by the Mexican television company Televisa for cable systems in the United States and relayed live to Europe.

Eutelsat 1F5

RAI Uno

The Italian public broadcaster RAI (Radiotelevisione Italiana) distributes its first channel, RAI Uno, in PAL unscrambled for 18 hours daily.

RAI Due

The Italian public broadcaster RAI distributes its second channel, RAI Due in PAL unscrambled for 11 hours daily.

NRK

The Norwegian public broadcaster (Norsk Rikskringkasting) distributes its first television channel (intended particularly for Norwegians living in remote locations away from terrestrial transmitters) in C-MAC unscrambled for 8–9 hours daily.

SIP (Service Information aux Parieurs)

SIP distributes coverage of horse races to offices of the Paris Mutuel Union (PMU), the French betting service, between 1300–1800 and 2000–2300. The signals are transmitted in B-MAC, scrambled.

TVE 1

The Spanish public broadcaster (Television Española) transmits its first channel in PAL unscrambled for 14 hours daily.

PanAmSat 1F1 45 0 °W

PanAmSat is a satellite using frequencies and orbital location designated for North American services, but this orbital location permits it to 'bridge' the Atlantic. Regulatory bargaining between the UK and the USA opened access to the US telecoms market for the UK company Cable and Wireless. In reciprocity PanAmSat achieved access to Europe. The sole television service currently carried on PanAmSat is Galavision (see Eutelsat 1F4).

Third generation, Direct Broadcast Satellites

Astra 1A

The first Astra satellite, and the first European satellite launched and operated by a private company, Société Européene des Satellites, was launched on 11 December 1988. Programme services from Astra began on 5 February 1989. The 16 channel medium power satellite, though designated as a telecommunication rather than a broadcasting satellite, has made possible direct-to-home broadcasting in Western Europe with transmission power at least double (and sometimes quadruple) that delivered by low powered 'second generation' television satellites.

Programming on Astra builds on, sometimes replicates, channels already established on other West European television satellites. Programme streams with familiar titles such as Screensport, Sky Channel and Lifestyle are transmitted from Astra for longer periods than they have been from low powered satellites, and new services have been, or are to be, established to complement existing ones.

An important feature of Astra is the satellite's capacity to transmit more sound than image signals. There is, therefore, an enhanced (relative to what is possible with second generation satellites) possibility of addressing different linguistic markets with the same programme by transmitting an image track simultaneously with sound tracks in, for example, Dutch, English and German. Indeed satellites' capability of delivering sound as well as images has been utilized to extend the reception area of radio services as well as television. Though satellite television has overshadowed satellite radio, Eutelsat satellites are now used (and there are also proposals to use Astra and the French DBS TDF1 for radio) for radio transmissions, including the BBC's World and External services.

Screensport and Lifestyle are transmitted from Astra with English, French and German sound tracks (Spanish will be added); FilmNet is transmitted with Dutch, French and Nordic language tracks.

For details of Sky Channel, Lifestyle and Screensport, ScanSat, Kindernet and MTV services on Astra see descriptions for each of these services under the heads of Intelsat VAF11, Eutelsat 1F4.

Three new services, unessayed on low powered satellites, have been established (or are proposed) for Astra as components of the Sky Television cluster of services. These are Sky News, Sky Movies and the Disney Channel.

On Astra Sky Channel becomes the core of a cluster of complementary programme streams, not a stand-alone channel. Sky Television on Astra offers: Sky Channel, Eurosport, Sky News and Sky Movies, and has proposed to add two further programme streams in late 1989: The Disney Channel and Sky Arts.

Sky Movies will be sold as a subscription package (bundled with the Disney Channel) for £12 per month per subscriber to TVRO owners and for £6.50 per month per subscriber to cable networks. Sky Television also charges cable networks 15p per month per subscriber for each basic channel redistributed by cable (though TVRO owners pay nothing for these channels). Cable networks are likely to retail the package to final consumers at a higher rate than the £7.10 (£6.50+15p+15p+15p+15p) demanded of them by Sky Television.

The Sky Movies channel was screened unscrambled for a preview promotional period until late 1989 (pending development of the encrypt/decrypt smart card system proposed, of which the capital cost to the viewer is estimated at £50). The channel screens films from 1600–2400 and proposes to extend the transmission period to 0200 and, when encrypted, to screen films for 18 hrs daily, offering *ca* 350 films per annum. The estimated annual programming cost of Sky Movies is £35–40 million per annum (*NMM*, 7.1.1989, p. 3).

The Disney Channel is proposed as a joint venture between Disney and News Corporation; costs (including transponder rental) and revenues are to be divided between the partners. The channel is to be screened for 18 hrs daily and will carry a mixed range of programming orientated to both adults and children. However the future of the Disney Channel is uncertain following Sky Television's law suit against Disney which was initiated in May 1989.

Sky News is a 24-hour news channel. It transmits an updated bulletin, lasting 30 minutes, every hour on the hour. The remainder of the programming time is claimed by Sky Television in its promotional material to be filled with: 'an exciting mix of informative and entertaining programmes, ranging from hard-hitting discussion shows and business reports to the latest showbiz gossip, health news and documentary series'.

Sky News takes feeds from ABC-World Television News, CBS and NBC-Visnews. It also retransmits NBC's US network news programmes, and a weekly review from the *Wall Street Journal*. Sky News has an estimated annual programme budget of £35 million per annum (*NMM*, 7.1.1989, p. 3).

The W H Smith channels, Lifestyle and Screensport, are transmitted for longer periods from Astra than they were from their low powered Intelsat carrier. Kindernet is to share an Astra transponder with Lifestyle.

Société Européene des Satellites (SES), the Luxembourg company that owns and operates Astra, has so far rented out 12 of the 16 channels on the first Astra satellite.

Two transponders are allocated to ScanSat (only one of which is currently in use), two to W H Smith television for Lifestyle and Screensport/Sport Kanal/TV Sport, one to MTV, six to Sky Television, for Sky Channel, Eurosport, Arts Channel (Sky Arts), Sky News, Sky Movies and the Disney Channel. The four other transponders, earmarked for German language services, are vacant. RTL Plus may possibly adopt one of these vacant transponders but no other German language services have yet indicated interest. However the success in securing English/Scandinavian channels for Astra 1A has prompted SES to advance plans for launching a second Astra satellite which would permit a total of 32 channels to be offered from satellites occupying a single orbital position.

Astra represents the first technically successful and adequately programmed 'third generation' satellite to be established in service in Western Europe.

BSB

The satellite of British Satellite Broadcasting (BSB) has yet to be launched. Initially scheduled for launch in September 1989 (and widely advertised to begin operation then) inception of BSB's services has been postponed to early 1990. BSB is to be a considerably more powerful satellite than Astra, with transponder transmitter power rated at 100 watts per channel, if only three channels are in use, and rather less if five channels are transmitted simultaneously.

BSB has yet to raise the £400 million necessary to the continuation of its enterprise. This capitalization of BSB's next round of financing may be more than usually difficult since the largest shareholder in BSB, the Bond Corporation, has had its credit rating reduced to the CCC level (indicating that loans to a company with such a rating are highly speculative). And Bond has been required by the Australian Broadcasting Tribunal to divest itself of one of its principal revenue generating assets – its Australian television stations. Even without the Bond Corporation's troubles, financing BSB will present very considerable difficulties. As BSB itself states: 'BSB is the only entirely privately financed DBS project in Western Europe. It will be one of the largest private sector start up investments ever undertaken in the United Kingdom.' Indeed the BSB project is the largest capital project (with the exception of the Channel Tunnel) currently under way in the UK.

BSB's first business plan assumed a monopoly of UK DBS services. The Sky/Astra initiative removed this potential privilege and BSB has also had to contend with the Government's proposals for radical changes to the terrestrial broadcasting order in the UK. These changes cannot with certainty be predicted before publication of the anticipated 1989 Broadcasting Bill, but it seems very likely that they will include increased supply of terrestrial television advertising time and/or establishment of terrestrial subscription television. Such measures would significantly reduce the likelihood of financial success for BSB, because they would mean that the two revenue streams on which BSB must draw for funding will be the locus of augmented and particularly intense competition.

Moreover BSB has experienced serious technical difficulties in getting its receiver apparatus into production. Rumours abound that the microchips required for its receiver do not work satisfactorily and that manufacturing difficulties also obtain with the distinctive rectangular receiving antenna, the 'squarial', that BSB has made its trademark.

The owners and investors in BSB at Spring 1989 were: Anglia Television, 3.3 per cent; Bond Corporation, 35.8 per cent; Chargeurs SA, 12.0 per cent; Granada Group 14.1 per cent; Invest International Holdings, 1.8 per cent; London Merchant Securities, 3.7 per cent; Next, 5.0 per cent; Pearson, 13.6 per cent; Reed International, 10.0 per cent; Trinity International, 0.7 per cent. Total investment committed at Spring 1989 was £353.5 million. BSB proposes four programme streams across the three channels for which it is currently licensed.

For each of its programme streams BSB proposes commissioning and scheduling original programming (*eg* it has commissioned Scottish Television to produce 260 entertainment shows for £2.5 million).

Should BSB succeed in its application to be licensed by the IBA for the fourth and fifth UK DBS channels these channels will be programmed as a News/Current Affairs channel (releasing NOW/The Sports Channel to be an exclusively sport channel), and as a music channel, to be named 'The Power Station' and run in conjunction with Virgin.

The Movie Channel

The movie channel is (like Canal Plus) to mix encrypted with unscrambled transmissions. It proposes free access during the daytime and subscription access 'at less than £10 per month' from 6 pm onwards. The Movie Channel promises at least two dozen new films per month and that most of its films will be shown within a year and a half of their theatrical release. BSB has secured access to the product of Paramount, MCA/Universal, Columbia, Warner and Cannon

NOW/The Sports Channel

The sports channel is to be largely unencrypted (though premium events will be encrypted and available only on a pay basis) and will be cross scheduled with a news programme stream. News programming is to be contracted to Crown Communications, and sport to the Mark McCormack TWI company.

Galaxy

Galaxy will be orientated at different times to different target audiences: to children, women at home and in the evening to young adults. It will be funded by advertising and programmed by subcontractors: New Media Television (the principal shareholders of which were Yorkshire Television and the *Daily Mail*) and Noel Gay television.

6 The future of satellite television in Europe

Satellite television in Western Europe promised, or threatened, to deliver three things: more choice in programming to audiences, competition to established terrestrial broadcasters and an internationalization of the audience. The history of satellite television in Western Europe is one of failure to deliver such changes.

Choice

Satellites *have* delivered new schedules of programming. Instead of the mixed programme streams characteristic of terrestrial television in Western Europe (a product of the hegemonic notion of public television as a service to a series of minorities, separate 'publics', each requiring a different diet of programming) satellite television has offered thematic programme streams, such as The Children's Channel, Eurosport and FilmNet. Choice has been extended both by such scheduling innovations and by increasing the number of competing programme streams. Some of these, such as RTL Plus and Sky Channel, have offered an increase in the number of mixed programme diets available to viewers. Others, such as Eins Plus and RAI Uno, have exactly or effectively replicated the programming of terrestrial public broadcasters.

This extension of choice delivered by second generation satellites is however a qualified one. The number of channels presently established is not economically sustainable. Probably all are losing money, and those few channels which anticipate a move into the black in the imminent future (subscription film channels such as FilmNet and Premiere and advertising funded general programme streams such as Sat Eins and perhaps RTL Plus) approach a positive cash flow in particular and not necessarily sustainable circumstances; as was shown by the demise of Premiere once the cost of programming was driven up by the entry of Sky Movies to the subscription film channel market. The future profitability of the most favourably positioned established services is conditional on there being no significant future change in their circumstances. Of these the most important is the status quo in terrestrial television, which is in fact very unlikely to be maintained. Even in the existing situation, no channels have yet achieved profitability.

Moreover the extension of viewer choice delivered by second generation satellite television is in large part an extension of viewer access to the film and television archives of the world: little original programming (with the qualified exception of news, sport and some low budget children's programming) has been delivered by Western Europe's television satellites.

The extent to which audiences have availed themselves of such extension of choice as has been offered by satellite television differs from market to market, largely influenced by the strength of the competition offered by other – particularly national terrestrial – television services. The nature of this differs from national market to national market.

Competition

Satellite television has potential competitive advantages relative to terrestrial television. It can deliver audiences across a wide geographical area; it can transmit signals to locations which are costly to reach using terrestrial methods; it is able to deliver significantly more channels (and television requiring greater bandwidth for signal carriage such as HDTV), than can terrestrial television, because the frequencies used for satellite television are relatively abundant. But nowhere has the potentiality for satellite television to deliver bandwidth-hungry signals been exploited (and this asset currently possessed by satellite television may decline in value as signal compression techniques reduce the spectrum space required for television).

Should HDTV become established, this competitive advantage of satellite television may become of decisive importance. For it is likely that Western European viewers will only be able to enjoy multi-channel HDTV if delivered by satellite (or MVDS) because many cable networks, and the frequencies available for conventional terrestrial television, lack the bandwidth capacity to deliver such signals.

Satellite television has successfully exploited its competitive advantages and ability to extend audience choice in two important West European markets – in Scandinavia and in West Germany. These markets share a common experience; their terrestrial services have been very strongly shaped by the ethos of public service broadcasting without a prior experience of the countervailing force of competition. In Scandinavia terrestrial television has only recently accepted advertising. On West German terrestrial television advertising is confined to blocks between programmes (not spots within programmes) and prohibited on the high viewing days of religious holidays and Sundays. Moreover, in both West Germany and Scandinavia public broadcasting has undersupplied audiences with entertainment programming. Hence satellite television in these two regions has provided a sorely-needed advertising medium, with consequent revenues sufficient to fund a diet of attractive programming. However these conditions, related to the orientation and funding of terrestrial public services, that have for satellite television opened a window of opportunity in Germany and Scandinavia, are not replicated in other European markets. And were terrestrial television in Germany and Scandinavia to change its programme mix and scheduling (as did public television in the Netherlands, resulting in a decline in consumption of satellite services) then even there satellite television would be very vulnerable to loss of audiences and advertising revenue.

In competition between terrestrial and satellite television, the former enjoys enormous advantages. It is established in the market place and is understood by viewers. It is both cheaper to receive and more generously funded than satellite television. The cost/benefit calculation performed (implicitly or explicitly) by viewers favours terrestrial services, as long as the programming offered by terrestrial services is programming of a type desired by viewers.

The history of second generation satellite television suggests that the potential competitive advantage of satellite television, its ability to deliver transnational audiences, is in practice not very significant. To date no programme stream, whether thematic or general, has built a transnational audience of sufficient size to deliver returns commensurate with costs.

Internationalization

Although channels which attempted to establish a transnational audience-funding programming by delivering a transnational advertising medium (pre-eminently Sky Channel and Super Channel), have either abandoned the attempt or have been bankrupted, there has been a significant internationalization of television as a consequence of satellite broadcasting. But the internationalization has not been, as was anticipated, through creation of an international audience of any significance. Rather, it has been through the establishment of broadcasting services serving a particular national (or linguistic) market from outside the jurisdiction of the governments of such markets. National development of broadcasting (whether terrestrial or satellite) in Western Europe is now impossible. For everywhere competing services are available, actually or potentially, from outside. The propagation characteristics of Herzian waves; the inability of engineers to tailor satellite footprints to national boundaries; the commitment of the European Commission to an integrated European market in goods and services; the potential commercial rewards from breaching national broadcasting monopolies (or quasi monopolies) and creating transnational audiences; all these factors combine to effectively dissolve the authority of national broadcasting institutions.

But this dissolution of national governmental authority does not necessarily mean that 'anything goes' in satellite television and that the airwaves are likely to be polluted in a war of programming governed by Gresham's law in which bad programming drives out good. The alarm voiced in some UK quarters about the proximate introduction into the UK of pornography by satellite is unlikely to be vindicated. Any sustainable programme service requires funding; for erotic/pornographic programming such funding may derive only from advertising or subscription. Advertisers have shown themselves unwilling to fund downmarket programming and have withdrawn advertising from popular media that over-egged the erotic cake. It is therefore unlikely that advertising will ever financially sustain pornographic television.

A subscription-funded pornographic service is theoretically possible. But in order to inhibit 'free riding' by viewers who have not paid a subscription, such a service would require to be scrambled. Prospective viewers would require a decoder and would be required to send subscriptions to the programme service provider. Governments retain ample powers to inhibit the sale (or rental) of decoders and to successfully obstruct transmission of subscription to service providers. It is unlikely, therefore, that a viable subscription-financed erotic service would be sustainable in the face of government opposition.

The sole possibility that therefore remains of access to pornography via satellite is that of 'eavesdropping' on services tolerated (and sustained by subscription) in neighbouring countries. Some UK viewers are thought to eavesdrop in this way on FilmNet's erotica. But in order to do so they require a FilmNet decoder which is not easy to obtain lawfully in the UK.

It is questionable whether it is a desirable aim of public policy in one EEC state to deny its citizens access to information which can lawfully be obtained in another. Has any damage accrued to the UK polity as a consequence of its citizens' access to spectacles such as those that may be seen on Canal Plus or FilmNet by UK citizens should they stay in hotels in Paris or Amsterdam where guests have access to such services on the television in their bedrooms? To take another example, has bullfighting developed as a sub-culture in the UK among those who have attended a corrida when in Spain? In this respect, as in others, the effect of satellite television has been far less than was anticipated.

Oligopolies

Although delivering augmented competition between programme streams the second generation of satellite television has delivered scant competition between established and new entrants to the media markets of Western Europe. This is perhaps unsurprising in a product/service field that has characteristically delivered loss rather than profit. For it is only well-established enterprises that can fund losses from other revenue streams, can reduce losses by vertically integrating activities and maximizing synergies between complementary product and service ranges, and have the expertise to take on risky ventures such as satellite television broadcasting. The world's three largest media conglomerates, Bertlesmann (RTL Plus), News Corporation (Sky Channel), and ABC (Screensport) are all actors in the European satellite television market along with other giants such as British Telecom (Children's Channel, Premiere, MTV), Fininvest (Tele 5, La Cinq), Springer (Sat Eins, Teleclub), Maxwell Communications (MTV), Pearson (BSB), Reed International (BSB) and relatively smaller media enterprises such as the UK ITV companies and West German and French press enterprises.

British Telecom and News Corporation offer convenient examples of the ways in which concentration of ownership may reduce real, if not paper, losses, and therefore of the dynamics that lead to oligopoly in this sector of the information economy.

British Telecom International (BTI) is a sub-contractor for the rental of transponders on a number of satellites. If BTI has surplus transponder capacity (possibly within an overall envelope of profitability) and sells such capacity to a subsidiary company, or a company in which its parent, British Telecom, has an interest (such as the Children's Channel) then any losses made by the subsidiary may, at least in part, be compensated for by the revenues paid to the service provider, in this case BTI. Such an arrangement may simply attribute as a loss to one enterprise what would otherwise be an invisibly under-used factor of production in another. A similar hypothetical relationship can be illustrated using the example of the bundle of companies which cluster under the News Corporation umbrella. Publicity, promotion and circulation of programme schedules for Sky Channel can be performed by News Corporation's UK press interests – the *Sun*, the *Sunday Times*, *Today*, *The News of the World*, *The Times* and *TV Guide*. *TV Guide* can itself be assisted in its entry on to the UK periodical market, and towards eventual profitability, by its preferential access to Sky Television's programme schedules and publicity. Programme supply to Sky Television's services can, and does, come from News' Corporation's interests in Fox Television and 20th Century Fox cinema films. There are, therefore, synergies and economies available to established media enterprises that are unavailable (or available only to a lesser extent) to new entrants to the market. All other things being equal (and particularly during a period of loss making) the market will tend to be populated by established rather than new enterprises.

Internationalization and the UK

The UK has long experienced competition from foreign radio stations (Radio Luxemburg and the pirates). And the proposed establishment of Radio Tara in the Irish Republic has created an agenda to which UK broadcasters and regulators have had to respond. Were the proposed UK national commercial radio networks not to be established, Radio Tara would uniquely offer a UK national radio audience to advertisers. But UK Television has hitherto been a UK monopoly. Except in Northern Ireland where RTE is received, there has been little UK reception of foreign television. Rather the flows have been in the reverse direction: viewers in Ireland, in the east via off-air reception, elsewhere via cable, have consumed UK television. BBC services are relayed via microwave and cable to viewers in the Low Countries. (In 1985 the BBC received about £1.5 million in rights income from Belgian cable networks.)

Second generation satellites had a nugatory effect on the viewing behaviour of the UK television audience, but the establishment of third generation, DBS, services on the Luxembourg Astra satellite has already significantly affected the development of UK television. The prospects for the UK-licensed DBS service proposed by British Satellite Broadcasting have been worsened, BSB's (and UK terrestrial television's) access to programming has been restricted and the costs of programming have been bid up. However the most important force of internationalization exerted by satellite television that has concerned the UK has not been an external force bearing on the UK from outside but rather has been UK-originated services bearing on the television systems of other European states. The UK has export rather than import as its primary role in the West European satellite television market.

London: a capital city of television

The United Kingdom has become, with Luxembourg, the preferred state of domicile for Western European satellite channels. MTV, CNN, Lifestyle, Screensport and Super Channel all originate from the UK and are directed towards wide European audiences. So channels too have been targeted to single national markets such as Canal 10, which is transmitted to Spain, and ScanSat, targeted on the Nordic countries. The UK as a domicile offers several important advantages to satellite television enterprises.

(1) London is a world centre for the international film and TV programme trade and is therefore a convenient location to acquire rights and to license programmes.

(2) The UK uniquely offers a competitive and market orientated telecommunications regime which does not confine channel operators to a monopoly PTT for the uplinking of signals to the satellite.

(3) Satellite television channels have to conform to the regulatory requirements of their state of domicile, the location from which signals are uplinked. In spite of the establishment of a new censorship body, the Broadcasting Standards Council the UK has a more permissive regulatory regime, especially in relation to television advertising, than have most Western European states. Sky Channel, for example, when aiming for a transnational audience, conformed to the advertising regulations of the UK Cable Authority (which in turn derive from those of the IBA). These regulations are less onerous than those of other European states (with the possible exceptions of Italy and Luxembourg).

However, the relations and structures evident in the second generation of European satellite television promise to be redrawn as the third generation begins. This new generation promises a focus on national (or, strictly, single-language) markets rather than the transnational market essayed during the second generation, and to deliver a period of hot competition to the television market in the UK.

The third generation: direct Broadcasting Satellites and the UK

Two clusters of DBS programme streams are likely to be established to serve the UK in the 1990s: one stream on the BSB satellite allocated to the UK's DBS orbital slot (and licensed by the Independent Broadcasting Authority), the other on the Astra satellite occupying the orbital slot allocated to Luxembourg.

The existence of the Astra services testifies to the rapid erosion of national communication sovereignty in the satellite era. For the UK licensed stream, that of British Satellite Broadcasting (BSB), was awarded a *monopoly* of UK DBS services and licenced for fifteen years (to 2004). Not only does BSB not now enjoy a monopoly but competing services, from Astra, have been established in the UK market before BSB's satellite has even been launched.

The first cluster of services began from SES's Astra satellite in February 1989, with Sky Television's initial four programme streams. These were followed on 1 March 1989 by the W H Smith channels Lifestyle and Screensport, and by MTV. Completion of Sky's projected cluster of programme streams is anticipated in late 1989 (see discussion of Astra above).

The second cluster, from BSB, British Satellite Broadcasting, is scheduled to begin in 1990 after the rescheduled launch of BSB's satellite, once to have been on 10 August 1989, now on an unspecified date in early 1990. There remains considerable doubt as to whether BSB will succeed in raising the capital it requires and whether BSB will ever enter service.

A third cluster of services may develop distinct from Sky and BSB through use of the two unallocated transponders on BSB's satellite. These transponders are the subject of competitive franchise allocation by the Independent Broadcasting Authority (IBA). BSB has submitted an application to be licensed to operate the two unallocated transponders, so it is possible that BSB will therefore control five programme streams rather than just three. In addition, further direct-to-home channels accessible to UK viewers, if established, are likely to be located either on the Irish DBS (at the same location as BSB's satellite) or on a proposed second Astra satellite (at the same location as the first).

Competition will develop between satellites, both to establish which will be the 'hot bird' for the UK (the satellite at which most DBS antennae will be pointed), and between different programme streams (whether carried either on the same or different satellites). In the first instance there is a strong complementarity of interests between programme streams carried on the same satellite, all wishing their satellite to become the 'hot bird' delivering the most attractive bundle of services. Hence BSB's welcome to the IBA's decision to bring forward advertisement and allocation of the remaining unprogrammed transponders on the BSB satellite. Even though competing services may become established on these transponders, BSB judges that to be a preferable outcome, rather than there being only a limited programme offer on its satellite. Similarly Sky Television has begun its services on Astra with its Sky Movies channel unencrypted (destined to be scrambled and available only to subscribers) in order to attract viewers to obtain dishes and expand the Astra viewing population as rapidly as possible, 'pulling' demand with as attractive and inexpensive a bundle of services as possible, even though later-date viewers will have to be denied free access to the channel.

Astra has established an important competitive edge by entering the market first, and by capturing the first (albeit emaciated) generation of dish/receiver purchasers and renters. Astra is likely to retain a receiving apparatus price advantage over BSB due to its use of the PAL encoding standard. BSB is likely to emphasize its anticipated competitive advantages over Astra which it judges to inhere in the higher power of its satellite permitting a smaller antenna and being less subject to signal deterioration due to adverse weather conditions, and use of the D-MAC signal coding standard as an evolutionary step towards HDTV.

It is possible that Astra/Sky's early entry into the UK DBS market has effectively scuppered BSB's initiative. Even without the financial difficulties of the Bond Corporation and technical difficulties of getting its receivers into production, BSB will find financing the further development of its service very difficult in the face of an unanticipated and firmly established competitor. However, should BSB establish its service the main arena of competition between the rival DBS channels will be in programming content and scheduling. The result of this competition will decide not only the balance of advantage between the two satellites, and between programme streams carried on the same satellite, *eg* between W H Smith's Screensport and Sky Television's (and the EBU's) Eurosport, but, most importantly, between satellite and terrestrial television. For satellite television programme streams will be competing not only among themselves but against terrestrial channels for audiences, advertising, subscription revenue and programming.

It is too soon to predict the results of the contests, but predictions are nonetheless being made. Citicorp Scrimgour Vickers (CSV), a UK stockbroker, predicts that BSB will 'win' with 1.9m dishes in 1991, *ie* 9 per cent of UK homes, against Astra achieving only 1.5 million or 7 per cent of homes. CSV estimate that BSB will generate £72.5 million in revenue in 1992 compared to Astra's £72.5 million. Of this income, most will be derived from subscription, and CSV predict 30 per cent of BSB viewers will subscribe to the BSB movie channel and 25 per cent of Sky viewers to its Sky Movies (*Broadcast*, 3.3.1989, p. 2).

CSV's estimates, though speculative (BSB had not launched its satellite or begun production of its 'squarial' at the time of writing), correspond to other forecasts and indicators. MORI (cited in the *Economist*, 21.1.1989, p. 29) found that 70 per cent of UK viewers polled would not pay for subscription television: CSV's estimate that 30 per cent of BSB and 25 per cent of Astra viewers will do so therefore corresponds to Mori's finding. CSV's estimate that aggregate penetration of DBS services is likely to amount to 16 per cent of homes by 1991 corresponds, roughly, to the uptake of cable subscriptions in UK homes passed by cable. But even on CSV's estimates it is clear that current expenditure (*eg* BSB's outlay on programming of £150 million per annum – only one of the costs of running a satellite television enterprise) will not be covered by income. The amortization of capital expenditure is still far away, and profit even further.

Clearly both Sky Television and BSB see their principal revenue streams coming from subscriptions to film channels. The examples of HBO (Home Box Office) in the United States and Canal Plus in France have engendered hopes that comparable revenue streams may be generated from subscription financed film channels in the UK. The Home Office subscription television study (1987) suggested an unsatisfied demand among UK viewers for 'premium' programming. Both Sky Movies and BSB plan to create a virtuous circle, whereby growing audiences finance new and attractive premium programming, which in turn delivers audiences to the complementary advertising-financed channels, which themselves become self-financing, thereby supporting the subscription-financed film channel and generating future high profit returns.

The preliminary sparring between Sky and BSB has centred on programming rights. Each enterprise has competed vigorously against the other for film properties, bidding up the price, both for satellite and for terrestrial broadcasters. Sky outbid ITV for rights to a hit US mini series *Lonesome Dove* paying a reported just under $200,000 an hour. An extraordinary sum when considered in relation to the tiny audiences Sky can achieve, and in relation to the price established by terrestrial broadcasters for comparable product (for example, the BBC paid £600,000 for the eight hour mini-series *The Thorn Birds* with which it achieved its highest audience in the 1984 season). BSB's competition with the BBC and ITV for rights to English league soccer established new record fees paid to the soccer authorities; £9 million per annum for coverage of mid-week matches. Both proposed UK DBS subscription film channels have attempted to ensure exclusive access to cinema film properties. BSB, in co-operation with the BBC, has acquired 481 first run films, 150 library films, 6 series, and 2 cartoons from MGM/United Artists for $100 million. The BBC is to have rights for terrestrial broadcasting two years after BSB and has paid under half of the total licence fee for such rights. Reputedly, BSB has a deal with Columbia which has secured 175 first run and 200 library films for $160 million, and another agreement with Universal/Paramount for rights to about 900 films (together with terrestrial rights worth about $60 million in resale value) for $300 million. Though BSB appears to have been more successful than Sky in securing access to film product, Sky has a unique immediate access to new films, thanks to the fact that its parent company, News Corporation, owns the 20th Century Fox company.

However for neither BSB or Sky is there a prospect of a positive cash flow, still less profit, in the foreseeable future. Sky's estimated expenditure in 1989 amounts to £115m (without including the costs of its offices and studios at Isleworth or its share of the costs of the Disney Channel or of Sky Arts). Sky's estimated receipts for its first year of operation on Astra amount to only £14.7 million (*NMM*, vol. 7, no. 1, 1989, p. 3). BSB has committed itself to expenditure, on programming alone, of £150 million in its first year of operation. High though these figures sound they are low in comparison to the programme budgets of terrestrial broadcast television. And attractive though the film properties are that Sky and BSB have acquired, some of their glamour vanishes as viewers realize that each film will be screened many times in each satellite broadcaster's schedule (BSB has rights to ten screenings for its Columbia films). To put the expenditure into proportion: Sky Television's programming budget of £115 million for 4 channels is approximately equivalent to Channel 4's budget for one channel. BSB's £150 million on programming in its first year yields an approximate average budget per hour of programming of £11,415 (assuming BSB transmits on only three channels for only 12 hours daily). This compares with average BBC programme expenditure per hour of £30,000 (statement by Michael Checkland, Director General of the BBC on BBC TV, January 1989). Even if the BBC's use of its resources is wasteful and BSB is able to achieve more bang per buck than can the BBC (assumptions that are unproved and untested) the BBC clearly enjoys a formidable resource advantage.

Terrestrial television has an established place in the market, and is able to deliver programming budgeted at considerably higher levels than can satellite television. Further, the cost of reception of terrestrial television is considerably lower than is the cost of reception of satellite television.

Costs of satellite television

	1989	1992
Astra 80cm Antenna or BSB 'squarial' and receiver	£259	£179
Decrypting device	£50	–
Remote control (optional)	£50	–
Installation	£80	£69
Total capital cost	£439	£248
Annual recurrent costs for subscription channels:		
BSB @ £9.99 per month (one channel)	£119.88	
Sky @ £12 per month (one channel)	£144	£144

Assuming the cost of television receiver and terrestrial antenna as 'sunk costs', the annual cost of terrestrial television to UK viewers is the cost of the colour television licence fee – £62 (in 1992 £77). The cost of satellite television is the cost of the licence fee *plus* the cost of additional receiving and decrypting equipment *plus* the cost of any subscriptions to pay channels. Indicative costs are shown in the table on the previous page, and are likely to be comparable for Astra and BSB.

Note that one-time capital costs indicated below are for one DBS system, and should be doubled if the hypothetical viewer considered is to receive signals from both BSB and Sky systems. Should Astra channels transmitted by W H Smith be required an additional decrypting device is necessary, incurring a commensurately higher capital cost. Depreciation is ignored (though industry sources estimate a 12 year equipment and satellite life cycle).

Clearly the costs of satellite television, for the consumer, are considerably in excess of the costs of terrestrial television. But satellite television promises to deliver programming costing, on average, *considerably less* per hour than does terrestrial television. Cost benefit comparisons between terrestrial and satellite television do not favour satellite television. Ted Turner, an influential and experienced commentator (and the owner of the most successful of United States' satellite delivered television 'super stations') stated at the industry meeting 'Television '89' that UK satellite television was headed for disaster and that: 'Both (*ie* BSB and Sky) will haemorrhage for a long time to come. A lot of fine people are going to lose a lot of money' (cited in *Broadcast*, 3.3.1989, p. 2).

Why, given the negative returns from established second generation satellite broadcasting and the evidently high costs and poor prospects of immediate return from third generation satellite television, are three clusters (BSB, Sky and the non-Sky Astra channels) of direct-to-home satellite television channels under way? There is a pervasive belief among investors in a pot of gold at the end of the rainbow. Gold that will reward the competitor tough enough and well enough capitalized to stand the very high losses that are likely to obtain whilst many players remain in the game. Gold is looked for from advertisers, attracted away from terrestrial television to a medium promising to deliver more finely differentiated and wealthier audiences than does ITV. Gold too is anticipated from subscription finance, for Canal Plus and HBO have demonstrated that substantial returns are achievable from subscription television.

However the UK broadcasting environment is not sufficiently like that of either France or the United States to render French or US experience comparable. Subscription finance for new television channels is likewise not a surefire way to fortune: in Canada subscription channels are remarkably unprofitable. Nor is the UK environment sufficiently stable to assure satellite television broadcasters that advertisers will migrate to satellite television and not to either new terrestrial services (Channel 5 or a Channel 4 selling its own advertising) or to a repositioned and reprogrammed ITV.

It is often suggested that satellite television will succeed in the UK market because video cassette recorders have done so. The UK VCR population *is high* in comparison to other western states. However the UK VCR population that rents (or purchases) pre-recorded video cassettes *is low* in terms of international comparisons. This suggests that VCRs are used in the UK to increase and/or restructure consumption of terrestrially broadcast television. That is, the cost/benefit balance offered by terrestrial television is preferred by viewers to that offered by cassette rental or purchase. It is true that VCRs and satellite television are not perfectly comparable. But consideration of the use to which VCRs are put by UK viewers suggests that there may be less uptake for satellite television than those who base their predictions on the penetration statistics for VCRs have hoped.

A better comparison is surely one based on penetration of cable among homes passed. The less than 20 per cent of homes passed by cable which in fact subscribe to cable suggests a less positive prognosis for satellite than some advocates have predicted.

But satellite television may offer a *different* and therefore more attractive diet of programmes than does, or can, terrestrial television. What are its chances if it offers programming unavailable via terrestrial television? Such programming can be of two kinds; programming purchaseable or producible in the open market, to which satellite television can enjoy preferential access, or programming unavailable to terrestrial broadcasters.

Satellite television can only enjoy long term preferential access to programming on the open market if it outbids terrestrial television in competition for it. Given the assured revenue stream enjoyed by the BBC from the licence fee and the established dominance of the advertising marketplace of ITV and Channel 4 (based on access to virtually 100 per cent of the UK population – a percentage to which satellite television cannot hope to approximate for decades) the prospect of satellite television consistently outbidding terrestrial broadcasters for programming seems unlikely to be realized. Unless, that is, satellite television commands a significant revenue stream from subscription. But the Home Office study (1987) suggested that there was no possibility of financing either BBC1 or BBC2 at funding and quality levels comparable to those delivered by licence fee finance, through subscription. That calculation must surely apply to other channels which will require funding comparable to BBC channels if they are to compete with the BBC for the attention of audiences. Admittedly Canal Plus has attracted impressive subscription revenues, but has done so with a programme mix including films that current public policy considerations would deny UK audiences; notably with pornographic films including the celebrated *Deep Throat*. Establishment of the Broadcasting Standards Council suggests that UK based satellite television channels will not be permitted to transmit such programming (which is similarly denied to terrestrial broadcasters). Provision of such programming from outside the UK will be conditional on adequate revenue streams being returned to the broadcaster from within the UK. Such revenue might come from either subscription or advertising finance. However it is inconceivable that advertisers would support such services sufficiently to render them profitable. Tesco supermarkets withdrew its advertising from the UK newspaper the *Daily Star* in reaction against its raunchy imagery, an action which precipitated a change in content of the newspaper. Moreover Sky Channel experienced difficulty in attracting advertisers to its highly rated show *Wrestlemania* because wrestling was perceived by advertisers to be undesirably 'downmarket'. So much more so will be pornography.

Subscription funding for pornography of foreign origin seems similarly unviable given the ability of the UK government to obstruct importation and sale of decoders and transfer of subscription funds overseas to the broadcaster originating the pornographic signals.

It seems therefore that satellite television will be unlikely to secure a consistently more attractive diet of programming than can terrestrial services (assuming that terrestrial services are not inhibited, either by regulatory action or their own values, from transmitting a programme diet that corresponds to audience tastes and interests). Since reception of satellite services will be more costly than reception of terrestrial services the cost/benefit balance of advantage decisively favours terrestrial broadcasters.

If terrestrial services remain substantially as strong as they are now, (perhaps stronger, depending on the content of the anticipated 1989 Broadcasting Act), and their programme schedule genuinely meets the needs and desires of audiences, then the prognosis for satellite television in the UK is very poor. We are unlikely to see successful completion of the UK's second most costly capital project, BSB. The poor prospects for satellite television are a powerful vindication of the UK government's refusal (in contrast to the governments of France and West Germany) to commit public funds to a greedy high technology fantasy.

7 The lessons from the second generation of satellite television

Two pervasive assumptions about New Communication Technologies and their consequences are well exemplified by Direct Broadcasting by Satellite (DBS). First, that the qualitative shift in social and economic structure from industrial society to information society through which, it is widely believed, we are now living is also a shift to a 'pay for' society where services, goods and relationships which were formerly free are now bought and sold.

Second, that the era of autonomous nation states controlling their own communication destiny has ended, and an era of globalization and international interdependence has begun.

Hitherto, most broadcasting services most of the time have been free to the user at the point of consumption. Now the future of DBS services chiefly depends on subscription funding, those who don't pay won't be able to consume. Until now most states most of the time have controlled their own communication destiny. To do so they collaborated successfully and designed broadcasting services to minimize trans-border spill over of signals. Now the venerable consensus which underpinned delicate European broadcasting ecologies has been replaced by aggressive colonization of hitherto sovereign airwaves from off shore bases. Mexico transmits television to Spain, Ireland radio to Britain and services from Luxembourg's Astra television satellites have captured millions of German and UK viewers.

An oversimplified picture perhaps but this is how the new era of Direct Broadcast by Satellite is often perceived. Yet certain, perhaps exacting, conditions require to be satisfied if this vision, or nightmare, is to be fulfilled. Most importantly the entrepreneurs offering new services must invent a product which satisfies final consumers – the television viewers – who consume not an innovatory technology but a service which, at the point of consumption, appears more or less like the familiar terrestrial services which have long dominated domestic leisure. Getting technology and regulation right are but pre-conditions to establishing and consolidating profitable services.

The relationship between satellite television and established services (notably terrestrial television and also cable and video) will decide the survival of direct to home satellite broadcasting. Clearly success will take different forms, for different markets have different terrestrial television services and different regimes of cable and video. No more than in clothing, food or motor vehicles can the characteristics of one market be assumed to be present in others. What works in Germany may not work in Britain, the cultural, social and political factors which have been decisive in shaping markets in Canada and the United States will not necessarily be present in France or Spain.

What DBS offers

The technology of Direct Broadcasting by Satellite has made possible entry to broadcasting markets in which access to new forces has formerly been restricted. Either for reasons of public policy (usually preservation and protection of established national broadcasters) or because of radio frequency spectrum shortages.

DBS enables entrants to television markets to maintain end-to-end control of the programme production, transmission and reception chain, whereas other 'monopoly busting' and 'frequency spectrum shortage abolishing' new communication technologies, such as cable and video-cassettes, do not. This is potentially important, for both cable network operators and cable programme providers have experienced (and continue to experience) problems due to lack of integrated end-to-end control of cable networks and programming. The quality of its programme offer ultimately determines whether or not a cable television service can successfully attract and maintain a subscriber base. Yet those responsible for programming, and advertisers who fund services do not always receive full and prompt intelligence from cable network operators about levels of uptake, or rejection, of a particular programme service. Divided responsibility gives cable network operators no less serious problems, for return on investment in capital intensive cable network ultimately depends on the attractiveness of programme services. Characteristically these are not controlled by the network operator. Moreover the 'culture' of companies engaged in building and operating networks has proven difficult to reconcile with that of companies concerned with scheduling and programming television channels.

Behind these initiatives are a combination of technological advances (notably more powerful launch rockets, a reduction in the size and weight of satellites, and increased sensitivity in receivers and increased transmitter power in satellites) that have made possible the delivery of numerous channels of television via satellite to final consumers at affordable cost. However terrestrial television is able to deliver a single channel, or a few channels of television, to large numbers of viewers (though not to viewers in remote and topographically difficult locations) at less cost than satellite transmission. And cable can deliver many more channels to viewers than can satellite television.

In broad brush terms DBS delivery of television is more costly than terrestrial delivery for a few channels and more costly than cable for many channels but for an intermediate number of channels it is likely to be the most cost effective system of service provision. The lower limit of the range in which satellite television is likely to be cost effective is bounded by the number of channels which can be transmitted by terrestrial transmission (in the UK, five). The upper limit is probably the number of channels which may be received with a single fixed receiving antenna or dish (here the limit so far set is that of the Astra system, which is to have three satellites in the same orbital slot, each satellite delivering sixteen channels). Thus within the range of six to forty eight channels DBS is an efficient method of television service delivery. At the time of writing UK viewers with

Astra receiving kit (and the appropriate decoders) were able to receive sixteen channels of television from the single Astra satellite so far commissioned (not all these channels are transmitted in English though they include the best known and most watched Sky Television channels). With additional equipment and a steerable antenna many more channels are available from other satellites. Listings magazines such as *Satellite Times* print programme schedules for seventeen channels in addition to the BSkyB channels.

However there are several intervening variables which have, and will continue to, significantly influenced the viability of DBS.

Whether direct broadcasting by satellite succeeds in one location and fails in another depends chiefly on the technological configuration of the service, the regulatory régime, the status of competing services and the corporate strategy of the satellite television service provider. Europe in the early 1990s has witnessed the successful establishment of a DBS delivery technology. And though satellite delivery of television has proven much less easy to regulate than has either terrestrial broadcasting or cable delivery of television European regulation has permitted establishment of DBS services (though not always in ways foreseen by regulators). But it has yet to be demonstrated whether DBS offers programming with sufficiently attractive trade-offs between costs and benefits to viewers in major markets, including the UK, for it to be financially viable in the long term.

The development of DBS in the UK, a story which has lasted barely two years, began as a corporate war of movement and now has the character of a war of position. The first phase of the war was an elimination contest between rival satellite services, the current phase is a battle of attrition between satellite television and its terrestrial competitors.

To the question 'what is a Direct Broadcast Satellite' a variety of 'correct' answers could be given. A common sense answer might run something like this. A Direct Broadcast Satellite (DBS) is a communication satellite which is sufficiently powerful to relay signals (usually television, sometimes radio and, in V-SAT format, telephony) transmitted from an earth station uplink directly to the homes of receivers who may be dispersed over a wide area. This process takes place without the use of intermediate transmission technologies such as cable or terrestrial microwave. Or a more formal, juridical, response might be: Direct Broadcast Satellites are communication satellites which conform to the definitions of the 1977 World Administrative Radio Conference (WARC). That is, a DBS is a satellite transmitting at a power of 65 dBW or more to receiving antennae (dishes) of 90 cm diameter. The latter definition is most helpful in understanding the international regulatory régime for satellites. The former for understanding why DBS in the UK does not fit the formal definitions on which regulators have relied. And why, in consequence, providers of Direct Broadcast Satellite television services have been able to 'end run' regulation and why consumers, viewers, have been able to 'end run' the market structure which operators have attempted to impose.

DBS technologies developed considerably between the 1977 WARC and the time of writing. DBS is now very different to what was foreseen in 1977. Signals (such as the news channel Sky News or the film channel FilmNet), transmitted from the Astra satellites occupying orbital slots allocated to Luxembourg and owned by SES (Société Européen des Satellites) are less powerful than those WARC envisaged (65 dBW) would be necessary for DBS. Yet these signals provide broadly satisfactory reception in most of

Great Britain to viewers using a 60 cm diameter dish, whereas WARC foresaw the need for a 90 cm dish. Advances in receiver and antenna design has made possible reasonably reliable reception of low powered satellite television signals at a considerably lower cost than was envisaged in the 1970s (although it is important to remember that reception of satellite signals is still considerably more costly than is reception of terrestrial broadcasting).

DBS is in more senses than one a 'frontier' communication technology. DBS is at a technological frontier but also at a legal frontier where the law is imperfectly adapted to prevailing conditions and where it is not, and cannot, always be enforced. Descriptions of DBS customarily resort to superlatives. Just as explorers' accounts of the geographical frontiers of the new world described fabulous benefits and fabulous hazards so too do contemporary accounts of DBS. Nowhere are the superlatives more abundant than in the market where the characteristics of DBS are best seen, the United Kingdom.

DBS in the United Kingdom

BSkyB (the major DBS service provider to the UK) is the UK's largest private sector investment project (bar the Channel Tunnel). The UK has more DBS homes than does any other single national market. UK DBS service providers are subsidiaries of some of the largest of transnational media corporations. Relative to other, earlier, new communication technologies such as cable television and lower powered satellite broadcasting, DBS has been strikingly successful in attracting customers. And DBS' financial losses continue at a very high level. At the time of writing (Spring, 1991) BSkyB had experienced difficulties in raising additional debt financing of £380 million and was to receive a further equity investment of £200 million from its owners, almost certainly bringing the total of accumulated funds invested in Direct-to-Home Broadcasting by satellite in the UK to more than £1.5 billion.

Although there are difficulties in defining a Direct Broadcast Satellite few would question that the UK based BSkyB (broadcasting to the UK and Ireland) is the world's most important DBS service. Yet although (in March 1991) there were approx 1.3 million UK homes receiving DTH (Direct to Home) signals and a further 300,000 receiving them via cable, considerably more DBS homes than there are in any other single country, the future of DBS television in the UK market remains in doubt.

In March 1991 BSkyB claimed to be in 2.3 million UK homes (BSkyB press release, 14.3.1991). However many cable homes are unable to receive all BSkyB channels due to limitations on the channel capacity of the cable system. The figure of 300,000 cable homes cited is an estimate of the number of UK cable homes with access to all BSkyB channels at this time.

BSkyB (and other DBS services directed to UK viewers) faces stiff competition and competes against mature technologies of service delivery (notably terrestrial broadcasting but also video-cassettes) which are well established in the market place. DBS should not be considered as a stand alone innovatory advanced technology but rather as one of several possible rival forms of television programme delivery.

BSkyB was established in November 1990 through the merger of two hitherto distinct

DBS services; British Satellite Broadcasting, BSB, (which began as the UK's licenced monopoly DBS service in 1990) and Sky Television (which began direct to home services in early 1989 from Luxembourg's Astra satellite. Sky had an earlier incarnation as a low powered, satellite to cable, service from 1982 to 1989). Prior to the merger Sky and BSB transmitted separate clusters of programmes. Sky screened four channels; news, sport, films and a general entertainment channel. BSB screened five channels; music, sport, films, arts/news and general entertainment. BSkyB's post merger service includes two film channels, a sport channel a news channel and a general entertainment channel. BSB, Sky and BSkyB were, and are, chiefly dependent for funding on revenues generated from their subscription film channels.

Not only are the levels of investment sunk in BSkyB substantial but so too have been BSkyB's losses. Sky's pre merger accumulated losses were estimated to be £350m (*Screen Digest*, Dec 1990, p. 270), and BSB's were perhaps £700m (*Cable and Satellite Europe*, Oct 1990, p. 8); the *Financial Times* estimate is even higher. On 23/4.3.1991 (p. 8) it referred to BSkyB's '£1.3 billion merger'. The losses continue; press comment on News Corporation's (the parent company of News International which was in turn the majority owner of Sky Television) financial results for the half year to end 1990 suggested that taking BSkyB off the News Corporation balance sheet for the last two months of 1990 had flattered News Corporation's results by Australian $20 million – about £8.3 million (*Financial Times*, 15.2.1991, p. 23). Although the merger between two rival satellite broadcasters (which formerly competed for the same UK market) has slowed the flood of red ink, the future of the new enterprise is far from assured. Kleinwort Benson estimates that BSkyB requires a subscriber base of 4.5 million homes in order to break even (*Satellite TV Finance*, 29 Nov 1990, p. 23). That is between a quarter and a fifth of UK television homes. This would be an intimidating target at any time. But particularly so during economic recession in the UK and when there is the prospect of an increase in the already stiff competition BSkyB faces from terrestrial television as a consequence of the planned establishment of a fifth UK terrestrial television channel (albeit Channel 5 will be accessible off air only to about 70 per cent of UK television homes).

The owners of British Sky Broadcasting, the company formed as a consequence of the November 1990 merger between the former British Satellite Broadcasting and Sky Television, are among the world's largest media corporations. 50 per cent of BSkyB is owned by BSB Holdings. BSB Holdings is, in turn, principally owned by large British press and television companies (and the French conglomerate Chargeurs). They include Pearson, the world's thirty-sixth largest information content enterprise (the publisher of the *Financial Times*, and with extensive media interests including Penguin Books and Longmans); Reed International, the world's twenty-first largest information content enterprise; and Granada (one of the 'big five' UK commercial television franchisees). The remainder of BSkyB is owned by Sky Television which in turn has News International as its majority shareholder. (News International is a subsidiary of News Corporation, the world's fourth largest information content enterprise, and is the publisher of the *Sun, The Times, The Sunday Times*, the *News of the World* and *Today*.) However the ownership of BSkyB gives a misleading impression of de facto control which rests with the minority partner, Sky Television. The majority of BSkyB management are former Sky executives, headed by the former managing director of Sky, Sam Chisholm, (formerly a chief executive of an Australian commercial television channel). The merger enabled each partner to reduce their exposure to risk and to cut their losses. But Sky Television has secured the best terms; 80 per cent of BSkyB losses between the merger date and any future break even point are to be borne by BSB Holdings. The first £400 million of profits are to be divided equally between the partners, the second £400 million are to be divided 80:20 in favour of BSB Holdings and thereafter profits are to be equally divided. However BSkyB's continuing difficulties in securing debt financing suggests that further capitalization of BSkyB is likely to require additional equity from BSkyB's owners. That being so it is possible that BSB Holdings will be able to secure revision in the partnership agreement since News Corporation already has very considerable levels of indebtedness and may be unable to contribute its share of new equity finance.

International comparisons

Compared to both cable television and a previous, less powerful, generation of television satellites, Direct Broadcasting by Satellite in the UK market has been strikingly successful in rapidly consolidating a viewer base of more than a million and a quarter homes. However the UK DBS market is not representative of other national markets (and, as has been previously stated, the ultimate viability of UK DBS cannot be taken for granted). Other major world television markets do not exhibit a comparable level of investment in and commitment to DBS. In the United States, for example, two DBS projects (including SkyCable which is partly owned by News Corporation) have experienced difficulty in raising investment funding and are yet to be established (although another service, K Prime Partners has initiated an experimental DBS service). France and Germany have launched very powerful Direct Broadcast Satellites but technical failures have cast a shadow over DBS in these major markets. (Problems with the French satellite's travelling wave tubes have reduced possible services by 40 per cent and a clamp inadvertently left on the first German satellite's solar panel array has rendered it useless for DBS service.) All the television channels currently carried by the German and French DBS duplicate terrestrial services (albeit satellite transmission has extended the reception area for these services). In contrast BSkyB's services are uniquely available via satellite. The technologically much less ambitious Marco Polo satellites launched by BSB, and the conservative Astra satellites used by Sky Television, have proved more reliable than the very powerful French and German DBS. (The French and German satellites transmit at 230 watts, Marco Polo at 110 watts, and Astra at 47 watts. Indeed, in terms of WARC definitions, Astra is insufficiently powerful to be classified as a true DBS).

The birth of BSkyB

The BSkyB merger eliminated destructive competition between BSB and Sky in the UK DBS market which was almost universally regarded as being too small to sustain two major competing broadcasters. Not only does BSkyB now dominate the UK DBS market (though it is important to recognize that viewers have access to numerous other non BSkyB DBS channels) but its bargaining power vis-à-vis customers, programme suppliers and cable redistribution networks has increased. BSkyB currently continues to transmit from BSB's Marco Polo satellite but is to terminate these transmissions and provide services exclusively from the Astra satellite. However BSkyB's position is neither uncontested nor cost-free.

Manufacturers and retailers (Comet, Philips, Nokia) of, now unsaleable BSB receivers and 'squarials' have sued BSkyB. The administrators of the Australian Bond Corporation, formerly the largest of BSB's shareholders, have also sued BSB Holdings, and BSkyB seems likely to be liable to pay compensation to BSB programme suppliers whose contracts have been cancelled. Moreover BSkyB has undertaken to replace former BSB viewers' equipment with Astra receivers. Other DBS television channels, notably those which, like Sky, are carried on Astra (such as MTV and the W H Smith Television channels Screensport and Lifestyle), though competing with BSkyB for advertising revenue, neither challenge BSkyB's dominant position in the UK DBS market nor are wholly disadvantageous to BSkyB. All DBS channels share an interest in the DBS receiver population growing and additional Astra programme channels are inducements to potential viewers to acquire satellite receiving equipment. On balance the interests of BSkyB are served by the existence of other Astra satellite television channels (so long as these channels do not compete for subscription revenues with BSkyB's central revenue source, its film channels) rather than by their absence.

Competition

BSkyB's principal competitor is terrestrial rather than satellite television. Direct Broadcast Satellite television has usually been considered as a stand-alone technology, with its own characteristics, its own regulatory structures, its own markets and its own corporate structures. But from the perspective of television viewers, the final consumers of DBS services, such an approach makes little sense. From the user point of view a Direct Broadcast Satellite is but one form of television service delivery. It competes for the two resources, time and money, which consumers expend on leisure activities with other forms of consumption.

Thus satellite television cannot be understood as an independent 'black box'. Rather it must be considered in relation to other, rival, systems of which terrestrial television is the most important. The video cassette sale and rental market is also a significant factor. Video cassettes and DBS film channels are partial substitutes for each other and the UK video market is particularly highly developed. It is more than three times larger than the next most important European video market (Spain). In 1990 UK video sale and rental revenues exceeded $US1.4 billion (*Kagan Euromedia Regulation*, No. 8, 20.12.1990, p. 2). The relative advantages of video, in comparison to DBS, include earlier access to new film titles, access to a greater range of films (including erotica) and the ability of viewers to better control their expenditure. Disadvantages of video are the inconvenience viewers incur in visiting the video outlet to choose and return their chosen title(s) and an often inferior sound and image quality to that of DBS. Elsewhere, notably in North America, cable television would demand consideration as an important rival to DBS. But in early 1991 the UK had fewer than a quarter as many cable homes as it had satellite television homes. However UK cable's chief selling point is as a delivery system for satellite television. And although cable prospects are currently prejudiced by the growth in the UK population of satellite receiving dishes future growth of cable networks may be decisive in determining the eventual viability of satellite television in the UK.

Television viewers are generally thought to subscribe to subscription services – on which the financial viability of BSkyB depends (for advertising revenues are so far nugatory) only if they are thereby able to achieve access to an increased supply of 'premium programming' – notably to feature films. There is no doubt that satellite television does offer increased access to movies beyond that offered by terrestrial television. In early 1991 a total of about 35 different films per week were screened on the UK's four terrestrial channels. However Sky Movies has showed an average of eleven feature films each *day*. Each week Sky Movies screens perhaps fifteen new films (each of which is repeated about five or six times. Over a month a Sky Movies viewer has had about 337 opportunities to view about 72 different films. Whereas terrestrial television in the UK screens considerably more than 72 films each month. However DBS has offered viewers earlier access to new films than does terrestrial television (though DBS is generally later in its film releases than either the cinema or videocassette markets). BSkyB has promised that its proposed two film channels will offer 24 different films a day and 750 films a year – each film will thus be screened about 12 times (BSkyB Press Release, 14.3.1991).

However, although Sky Movies (and the residual BSB Movie Channel) certainly now offers greater access to premium programming than do UK terrestrial services the rights to premium programming are costly and there is a world shortage of premium material. For the foreseeable future UK DBS channels will be less able to sustain high levels of expenditure on premium programming – principally films but also sport and popular music spectaculars – for the UK market than will terrestrial services. Subscribers to DBS film channels already express discontent to researchers and in letters to satellite listings magazines about the number of repeated films. Of eleven readers' letters published in *Satellite Times*, 1–15 March 1991, three concerned viewing of the, officially unavailable FilmNet channel, one the erotic game show *Tutti Frutti* (on RTL Plus) and three complained about advertising, repeats and type of programming on the Sky Movies channel. However the most important signal, the level of subscriber 'churn', seems to indicate relatively low levels of dissatisfaction. Although 'churn' and its extent is one of the most sensitive matters of commercial intelligence the indices which are publicly available suggest a relatively low churn rate of 10 per cent for BSkyB subscribers (*New Media Markets*, 22.11.1990, p. 11).

What are the relative cost benefit trade-offs between satellite and terrestrial television? Reception of satellite television is more costly than terrestrial television. It requires expenditure on receiver, antenna, decoder and installation (at current prices about £329) beyond the common costs (television set and receiving licence) incurred in reception of terrestrial and satellite television. The premium film channel, Sky Movies, costs a further subscription (currently £9.92 per month rising to £9.99 per month. BSkyB's second film channel 'The Movie Channel' will also cost £9.99 per month: the two channels will together be available to viewers for £14.99 per month). If satellite and terrestrial television were perfect substitutes for each other there would be no point in viewers paying for satellite television. The viability of satellite television therefore depends on it providing either distinctive programming and scheduling differentiating it from the terrestrial services available for consumption at lower cost, or on extending choice of viewing opportunities at any particular time.

In both respects DBS television in the UK has succeeded. It has acquired films, television programming and rights to sporting events (which were sought by terrestrial television) and denied them to viewers of terrestrial services. Sky Television (and latterly BSkyB), thanks to its position within the global News Corporation, has preferential access to the output of the Fox film studios and schedules one of the hottest of 1990's television shows *The Simpsons*. It has established a distinctive diet of programming which is uniquely available to satellite viewers. And its distinctive scheduling patterns with strongly 'branded' channels with regular programme junctions contrasts with the mixed programming and – news excepted – irregular programme junctions of terrestrial television. Satellite television accounts for about a third of viewing time in UK homes with access to it, and generally achieves a higher audience share in satellite television homes than do Channel 4 and BBC2. Such viewer behaviour presents terrestrial services, and ITV in particular, with an acute dilemma. To counter the attractions of satellite television, change in the scheduling and programme mix of terrestrial services is indicated, either in the schedules of particular channels and/or by 'branding' terrestrial channels so they have a more strongly marked product identity. Hypothetically BBC1 might be branded as a film/drama / sport channel and BBC2 as a news and information channel, but UK terrestrial services are constrained in their responses to the satellite challenge; Channel 4 and the BBC by Charter and statutory expectations; and ITV franchisees by their need, particularly at the time of writing, to satisfy their regulator, the ITC (Independent Television Commission), that they offer 'quality' programming. A track record for quality is likely to be a bankable asset in the forthcoming auctions for Channel 3 and 5 franchises. Counter programming to *The Simpsons* is unlikely to satisfy the ITC's criteria. And well established organizations, such as the UK terrestrial broadcasters, have a considerable weight of institutional inertia slowing their ability to change.

Satellite television offers viewers further benefits. Even if satellite television programming is not significantly different to terrestrial television (but is a close substitute for it), viewers may sufficiently value the extension of their viewing choice offered by satellite television at any particular time. UK DBS provides both extension of viewer choice and access to programming absent on terrestrial channels (albeit erotica is available only to 'eavesdroppers' on Dutch and German channels). The clusters of satellite channels, news, sport, films and general entertainment, offered by Sky Television; and BSB's music, films, sport, arts / news, and general entertainment channels (to be rationalized by BSkyB into a news, a sport, a general entertainment and two film channels) clearly extend viewers' choices at any particular time beyond those offered by terrestrial television. The viability of DBS in the UK depends on viewer assessment of the balance of costs and benefits which are associated with satellite television. DBS offers distinctive programming and an extension of choice, but at a cost. Its success is vulnerable to changed programming strategies by terrestrial services with deeper pockets, and to viewer behaviour which maximizes benefits whilst minimizing costs, notably through 'piracy'.

BSkyB, is experiencing an unregulated invasion of the UK market it hoped to dominate. The major funding stream for BSkyB, subscription income from its film channels, is compromised by a growing population of unauthorized de-coders which enable UK viewers to view the Belgian/Swedish film channel FilmNet for a one off capital payment of less than an annual subscription to Sky Movies. FilmNet also provides erotic programming which is not available from BSkyB or any UK broadcaster. *Satellite TV Finance* (29.11.1990, p. 19) estimated the UK population of unauthorized FilmNet decoders to be 100,000; earlier *Cable and Satellite Europe* (3.1990, p. 18) estimated a considerably higher penetration of 20 per cent of UK DBS households with a FilmNet decoder. Viewers with a taste for erotica but without decoders are able to watch the German RTL Plus channel's *Tutti Frutti* strip game show (screened in 3D for the first time in March 1991).

DBS in the UK exemplifies an intriguing paradox: the successful establishment of Sky Television (and its dominance in the BSkyB partnership) was due in no small part to its successful end-running of UK regulation by using a European satellite. However its main revenue stream – subscriptions to its film channels – is vulnerable to unregulated viewer 'piracy' of a European film channel, FilmNet.

Regulation

The history of satellite television in Western Europe has been a history of entrepreneurs using the new medium to end-run regulation. The 'second generation' of television satellites (low powered satellites on which television services were first established in the early 1980s) were 'telecommunication' rather than 'television satellites'. Regulators did not respond effectively to the use of frequencies and satellites reserved for telecommunications for distribution of television signals. The use of a satellite within the jurisdiction of one state to deliver television signals to viewers in a second state (signals which were sometimes uplinked from a third state) clearly rendered national regulation a virtually impotent instrument for the realization of public policy goals. Long established European broadcasting ecologies based on national television monopolies (or carefully limited competition) were rapidly overturned by satellite television.

In Sweden, for example, Sveriges Radio's monopoly was successively eroded by the English language signals of Sky and Super Channel, by the Nordic commercial channel TV3 ScanSat, and in 1990 by the Swedish satellite channel TV4. (In turn a regulatory initiative, to establish terrestrial commercial television in Sweden for the first time, threatens the commercial viability of satellite television services serving the Swedish market. (*Financial Times*, 7.3.91, p. 4) In Germany (formerly West Germany) the Luxembourg based satellite channel RTL Plus and the German Sat Eins, assisted by other satellite channels (Premiere, Pro 7, Teleclub, Tele 5 and Sportkanal), broke the monopoly of German public broadcasters. In the UK, the IBA's (the IBA, Independent Broadcasting Authority, was the predecessor of the current UK regulator of satellite television, the Independent Television Commission, ITC) planned establishment of direct to home satellite television (based on a monopoly licence to BSB) was forestalled by Sky Television which rapidly transformed itself from a low powered second generation service, largely distributed via cable and transmitting an advertising financed single programme channel to a multilingual Europe wide audience, to a multi-channel subscription funded service transmitted direct to homes in the UK and Ireland from a Luxembourg satellite.

Sky rapidly pre-empted BSB and established its four channels in a UK market place which had been reserved for BSB before BSB launched its satellite. By September 1990, a month before the merger of Sky and BSB into BSkyB (which was announced on 2 November 1990). Sky had secured a direct-to-home viewer population eleven times greater than had BSB; 962,000 to 88,500 (*New Media Markets*, 11.10.1990, p. 3).

The UK Broadcasting Act 1990 provides specific regulations for satellite television. It distinguishes between domestic and non-domestic satellite services. Domestic services are transmitted from and to the UK on frequencies reserved for satellite television. The Act identifies two kinds of non-domestic services; those which are not transmitted on an authorized (*ie* for television rather than for telecommunications) frequency but are uplinked from the UK and are designed for reception in the UK. And those which are transmitted from outside the UK, but are designed for reception in the UK (or elsewhere) *and* have programme elements provided from the UK by 'a person who is in a position to determine what is to be included in the service' (Para 43 (2)b).

Domestic satellite services and terrestrial Channel 3 services fall under the same licensing and regulatory régime. Licensing of a non-domestic service is conditional on the service satisfying regulatory requirements relating to programme content; impartiality, taste and decency, absence of politically provocative material, *etc*.

The 1990 Act is innovatory (in respect of satellite television) in its extension of UK jurisdiction to programme services which are transmitted from a non-UK satellite. The best known non-domestic satellite service was, and is, of course, Sky Television (now BSkyB); assembled in West London, uplinked from the UK and transmitted from the Luxembourg satellite Astra. However Sky is not the only non-domestic service established in the UK. The TV3 channel, which reaches *ca* 2.6 million Scandinavian households, is also uplinked from the UK to Astra by the Scansat company. The only UK domestic satellite service BSB (now subsumed into BSkyB), which was granted a monopoly licence by the Independent Broadcasting Authority to provide DBS services in the UK for fifteen years.

BSkyB occupies an anomalous position in terms of regulation. In respect of its broadcasts from the former BSB satellite, Marco Polo, it is the authorized UK DBS franchisee. Its transmissions to the 120,000 or so unfortunate purchasers of the BSB 'squarial' continue to conform to the EC regulation that DBS signals must be encoded in MAC standard. BSkyB's transmissions from Astra are encoded in PAL. Because Astra is technically not a television but a telecommunication satellite, programme services transmitted from Astra have been able to 'end-run' EC stipulations on use of the MAC standard, and UK regulations limiting concentration of media ownership. As the successor of BSB, BSkyB is under the protection and subject to the regulation of the ITC whilst it transmits from the Marco Polo satellites. (In many respects of course BSkyB no longer satisfies the conditions on which BSB was licenced: however in respect of its broadcasts from Astra – to an estimated 1.1 million dish owning UK homes – BSkyB escapes UK regulation.)

In fact BSkyB proposes to apply for a licence as a non-domestic broadcaster in order to secure access to UK cable networks (if successful BSkyB will incur a licence fee of up to £1.47 million). But here too, were regulatory requirements to become onerous, BSkyB could simply transfer its transmission and scheduling activities outside the UK. This option, albeit one which would restrict viewer access to BSkyB via cable (which may become the key medium, making or breaking BSkyB's ability to achieve the *ca* 4.5 million households required to achieve financial viability) gives BSkyB considerable bargaining power in any dispute with the ITC. The limits of ITC jurisdiction and BSkyB's bargaining power have enabled BSkyB to continue transmissions from the Marco Polo satellite in spite of *prima facie* breach of several provisions in Schedule 2 of the Broadcasting Act. (These concern, *inter alia*, the citizenship of a person with a controlling interest in a licensed broadcaster and measures to limit concentration of mass media ownership.)

Whether or not effective regulation of trans-border satellite television can be established it is clear that regulation of satellite television will be possible only on a basis of shared sovereignty between hitherto autonomous jurisdictions. In spite of Council of Europe and European Community initiatives (notably the Council of Europe's Transfrontier Television Convention 1989, and the European Commission's Television Without Frontiers Directive 1989) enforcement of regulations (whether national or transnational) remains a matter for national governments which oftimes lack effective instruments of enforcement and common regulatory interests.

The prospects for effective transnational regulation are not encouraging. For example the Government of Ireland has not taken action in respect of Radio Atlantic 252, formerly Radio Tara, (built and operated from the Republic of Ireland to reach listeners in the UK). A senior source in the UK Radio Authority wryly observed that the UK has more important matters than Radio Atlantic's entry into the UK radio market to pursue with the Government of Ireland. And there are no current instruments of redress against UK viewers of foreign satellite television services using unauthorized decoders although unauthorized reception of programmes provided from the United Kingdom is a criminal offence. But neither UK copyright law nor the 1990 Broadcasting Act enable a copyright holder to take civil action against a viewer, or listener, who receives a signal she or he is not authorized to receive (*eg* by using an unauthorized decoder). However the Broadcasting Act 1990 does contain provisions amending the Copyright, Designs and Patents Act 1988 (Section 297) by making sale, import, hire or manufacture – but not possession – of an unauthorized decoder of UK originated services unlawful. There are some uncertainties about the power of the relevant section in the Copyright Act (and the amendments to it effected by the 1990 Broadcasting Act) but the Act seems to extend protection only to services originating in the UK and in jurisdictions offering reciprocal protection. Thus there seems to be no civil or criminal offence committed by possessing, selling or manufacturing decoders to non-UK programme services. In the most significant case – unauthorized use of FilmNet decoders by UK viewers the principal loser, BSkyB, lacks standing and FilmNet has small incentive to take action. For FilmNet suffers little from unauthorized UK reception of its signal. It has no rights to screen its programming in the UK market and can generate no revenue in the UK. FilmNet therefore does not lose from unauthorized UK reception. That it scrambles its signal (and periodically changes its decoding protocols – though so far not sufficiently to constitute a serious obstacle to pirate decoder manufacturers) has been sufficient to satisfy the owners of rights to the films it screens; FilmNet has therefore no financial interest in initiating criminal prosecutions in the UK (so long as there is no significant leakage of unauthorized decoders into its prime markets). BSkyB, which does have a financial interest at stake (for FilmNet is a fairly close substitute for its subscription film channels), lacks standing to initiate prosecutions for the sale of unauthorized decoders of programme services other than its own. A walk along Tottenham Court Road or browse through the advertisements in *What Satellite* are eloquent testimony to the pervasive availability of unauthorized decoders.

Technical standards and electronics industry policy

Whatever the future of BSkyB, and whether or not it achieves a positive cash flow in time to secure its future and deliver an eventual profit to its financially troubled parents, BSkyB has had a decisive impact on the development of the West European Direct Broadcast Satellite television market. Sky Television's choice of proven technologies and conservative engineering systems has been vindicated; notably the low powered and conservatively engineered Astra delivery system and the PAL video encoding system. BSB, partly from choice but also compelled by regulatory necessity, used the novel MAC video encoding system. Although the BSB Marco Polo satellites have worked well (not least in comparison to the French and West German DBS) BSB's choice of a technologically innovatory antenna (which proved impossible to manufacture to schedule and to cost), the 'squarial', was unfortunate. BSB was also cursed by difficulties in developing, debugging and manufacturing receiver chips. These delays materially assisted Sky television in consolidating its 'first mover' advantages; particularly in building up a relatively large receiver population before BSB entered the market.

BSB's commercial failure and the technical difficulties experienced by the German and French DBS have severely compromised the future of the MAC standard on which many of the European Commission's hopes for a European electronics policy rested. The European Commission required use of MAC by European DBS services from 1986 to 1991. However Astra has used PAL rather than MAC standards and was able to end-run regulatory requirements because it is technically a telecommunication rather than a broadcasting satellite.

MAC (Multiplex Analogue Components) was developed by the Engineering Department of the UK Independent Broadcasting Authority, the French Government broadcasting and telecommunications engineering research establishment and Philips and Thomson. Three rationales were advanced for MAC. First, it offered superior picture quality to that available from established standards; notably PAL. Second, MAC would establish an incremental path towards a European High Definition Television (HDTV) system. And third, MAC patents would protect European television manufacturers from lower cost manufacturers in Japan and SE Asia in the same way that the, now defunct, PAL patents had done.

But although the Commission supported MAC, the benefits of MAC were evident only to those few BSB subscribers who possessed MAC receivers. Most BSB homes viewed MAC signals on PAL sets and did not experience a different quality of reception with MAC transmissions to those experienced with PAL transmissions. Latterly a fully digital route to European HDTV rather than the analogue MAC route has been gaining support. And the industry protection rationale for MAC has been challenged on the grounds that

MAC would not offer effective protection (and that the long-term interests of the companies and the public are not served by protection) and would raise the price of consumer electronics equipment to European households. BSB's receiving equipment – albeit for reasons that are not wholly explicable in terms of the additional cost of MAC compared to PAL components – was *ca* £100 more expensive than was the PAL based Sky Television equipment.

However the future of MAC, in spite of the Commission's first MAC Directive, is in doubt. The low cost of PAL components (because PAL patents have expired) assisted rapid growth of the Astra PAL based system. And important European terrestrial broadcasters favour PAL based systems rather than MAC for forthcoming generations of television equipment. The BBC advocates HD PAL, German public broadcasters PAL Plus and the UK ITC and ITVA (Independent Television Association) now support a system known as E PAL. But it is chiefly Astra's success which has compromised the Commission's plans to establish MAC as a general European video standard. Although MAC continues to be supported by France, *Cable and Satellite Europe* made the acerbic comment that there are only a 'few thousand D2-MAC units in use in the whole of Europe' (*Cable and Satellite Europe*, 1.1991, p. 44). The revised, second, Directive on satellite television transmission standards is unlikely to have sufficient clout to establish MAC as a general satellite television transmission standard in Western Europe.

The prognosis for UK satellite television

What we know about reception of DBS services in the UK is that households with access to DBS television do watch it. In Astra households in autumn 1990 terrestrial television was viewed for 70 per cent of viewing time and Sky Television services for 23–25 per cent of viewing time (*Satellite Television Finance*, based on BARB data, 29.11.1990, p. 21). Churn rates for film channels appear to be lower than those experienced for pay television services in North America, but higher than those experienced in France. However satellite television households are a minority of the total population of UK television households, and although the total number of satellite television households continues to grow the rate of growth seems to have slowed. The most important factor which will shape the future of DBS is likely to be the relative competitiveness of terrestrial services in the different national broadcasting markets entered by DBS services. The terrestrial services in the UK offer tough competition. They are relatively well funded and terrestrial broadcasters consistently enjoy considerably higher budgets per programme hour than do satellite broadcasters. Moreover terrestrial services can be received at considerably lower cost than can satellite programme services; and UK terrestrial broadcasters enjoy a well established market position.

Survey research suggests that the majority of UK TV households do not intend to become satellite households. 77 per cent of UK adults polled stated they did not propose to acquire a satellite receiving dish: 33 per cent of those who rejected satellite television did so because the 'BBC and ITV provide enough channels for them'; 25 per cent found satellite television 'too expensive'; 23 per cent said they do not watch much television; and 8 per cent that Sky and BSB programmes are 'poor' (*New Media Markets*, 22.11.1990, p. 11). This empirical research supports the analytical conclusion that cost-benefit comparisons favour terrestrial services over satellite television. Both suggest that if BSkyB is to reach the 4.5 million households Kleinwort Benson estimated to be required for break-even *every* household which did not respond negatively to the enquiry 'Will you acquire satellite receiving equipment?' *must* be recruited by BSkyB; achieving a viewer population of 4.5 million households will not be easy. Moreover, whilst viewers are able to access FilmNet for the cost of a decoder priced at less than the cost of an annual Sky Movies subscription, UK DBS households are unlikely to deliver the levels of subscription revenues on which the future of BSkyB will depend.

The acid test of viability of UK DBS is the level of household penetration achieved and the number of these households which subscribe, and maintain their subscriptions, to the film channels. Certainly the pace of household penetration appears to have slowed. In January 1990 receiving dishes were being installed at a rate of 75,000 per month. In January 1991 BSkyB claimed an installation rate of 34,000 per month but an independent survey by Continental Research suggested a rate of only 17,000 per month (*New Media Markets*, 14.2.1991, p. 1) although the total rose in February 1991 to an estimated 41,000 dishes per month (*Financial Times*, 11.3.1991, p. 6). However, even if a monthly installation rate of 40,000 is taken as a basis for calculation, annual growth in BSkyB households is only 2.5 per cent of UK television households. If the claimed rate of increase is maintained, break-even point (estimated at 4.5 million households) will take between five and six years to achieve. Moreover satellite television homes are concentrated predominantly (70 per cent) in social groups C2, D and E; not in the ABC1 groups most sought after by advertisers. Subscription funding thus seems likely to remain vital to direct-to-home satellite television funding. BSkyB claim 'about 1m' subscriptions to Sky Movies (BSkyB Press Office, 13.3.1991) but is not prepared to disclose churn levels.

These estimates are probably the best available in the public domain but they should be used with caution. They underscore how small a proportion of total UK television viewing satellite television represents. And, in consequence, how difficult it will be for UK DBS to achieve commercial viability. However UK commercial television is experiencing a period of turbulence and during the run up to new franchises its priority is likely to be satisfying the ITC's quality criteria rather than undercutting BSkyB. It is not surprising that Murdoch interests have argued for the BBC to change from a full service broadcaster to one which supplies programming for minorities ignored by commercial services.

Conclusion

The risks of launching a DBS service in the UK market are evident. Yet considerable resources have been committed to UK DBS services. What are the potential rewards that are commensurate with the high risks run by those who have sought established DBS television services in the UK?

DBS is a medium, like all broadcasting services, with high fixed costs but low marginal costs in reaching additional consumers. Growth in the subscriber base for a subscription channel feeds through rapidly to the bottom line. Once costs have been covered profits will grow in direct proportion to growth in the subscriber base. After the BSkyB merger the UK (and Irish) markets have no rival Pay-TV subscription services, so monopoly access to this potentially lucrative revenue stream is attractive (though as stated above competition from terrestrial television and video is intimidating). Moreover DBS services are less effectively regulated than are terrestrial services. A person who would, under the terms of the 1990 Broadcasting Act, be excluded from access to UK broadcasting, is able to enter that UK broadcasting market as a satellite broadcaster.

Examination of Direct Broadcasting by Satellite of television to the UK emphasizes the complex cross-impacts between different forms of television service delivery. Not only is the future of satellite television likely to depend on the balance of costs and benefits it offers relative to terrestrial television and if consumers of satellite television are not to be subject to monopolistic rises in service prices and falls in service quality, then terrestrial broadcasting must be able to compete effectively against satellite television for viewers. The terrestrial broadcasters begin with significant advantages. They are well established in the market place and provide virtually universal service, reception of terrestrial services requires a lower capital investment in receiving equipment, and services are free at the point of consumption. However terrestrial broadcasters are constrained, both by institutional tradition and by regulatory requirements, from adapting their service offer to compete against satellite services. Increasingly the mixed programming of the four UK terrestrial channels looks unfocussed in comparison to the 'branded' channels of satellite borne competition. Not only has satellite television been able to offer new programming and an extension of viewing choices but it has successfully introduced new forms of programme packaging and scheduling.

The configuration of DBS television services reflects the historical weakness of UK (and European) broadcasting regulation. To state this is not to suggest that the outcomes desired by regulators are to be preferred to those which have eventuated in the market place. But it is a matter of fact that the existence of BSkyB, and BSkyB's vulnerability to 'piracy', not of its own but FilmNet's pay channel, are products of a lack of regulatory power. However changes in the regulatory régime governing UK cable (notably the decision to permit cable networks to provide telecommunication services and enjoy assured rights of interconnection to trunk and international networks) is likely to lead to an acceleration in cable network construction and cable becoming a more attractive means of accessing satellite television signals. In turn an increased reliance on cable as a distribution medium is likely to render UK regulation of satellite services more effective. Increased regulatory effectiveness may assist in protecting consumer interests by ensuring a proper balance between service costs and quality, but is likely also to reduce UK television viewers' access to television programming enjoyed by other European Community citizens. Trans-border spill over of television signals via satellite has enabled UK television audiences, for the first time for most viewers, to view television from other European states and to experience varieties of programming not offered by UK broadcasters.

DBS has caused UK viewers to make cost benefit analyses (whether implicitly or explicitly) about their television consumption. They have had to balance the undoubted extension of choice of viewing opportunities and, to a lesser extent, greater variety in programming offered by DBS television against the increased costs of consumption which attach to satellite television. Many factors in this calculation favour terrestrial services. It will be interesting to see whether UK public service broadcasters are as slow to respond to changed competitive circumstances as have been their equivalents in other jurisdictions. It is still a small minority of UK television households which have purchased access to satellite services. These viewers have shown both that they are prepared to pay for access to new services and more viewing choices. However they have also shown the importance of low cost of service delivery – Sky's inexpensive Astra/PAL technology permitted rapid establishment of a viewer population, BSB's costlier and later MAC/Marco Polo configuration did not. Moreover the significant number of viewers who have taken the opportunity to reduce the cost of premium programming (and to secure access to erotica unscheduled by BSkyB) by purchasing an unauthorized FilmNet decoder underscores the importance viewers appear to attach to favourable cost-benefit relationships.

Too many factors are in play for predictions to be made with confidence about the future of DBS in the UK market. Satellite television has come a very long way in a very short time. In two years Sky Television has eliminated its authorized UK competitor and established a partnership with BSB on very favourable terms. It has rapidly built a viewer population of impressive size (though BSkyB is less forthcoming about the vital statistics of subscription and churn ratios in respect of its film channels). But the partners in BSkyB have incurred some horrifying losses. Sky Television's losses would be impressive in any circumstances other than the present ones where its sunk costs have been dwarfed by BSB's. An estimated investment by BSB and Sky Television of more than £1,000,000,000 in BSkyB has yielded a viewer population of perhaps a third of the size of that authoritatively estimated to be necessary for the service to break even. BSkyB, and DBS in the UK, has come a long way, but it still has a long way to go. The first million viewers are likely to have been easier to find than will the last million needed.

References

The data used in this study principally comes from secondary sources, notably the excellent UK trade journals *Broadcast, Television Business International, Cable and Satellite Europe* and *New Media Markets*. I am particularly indebted to Nick Snow of *Cable and Satellite Europe* for his permission to make extensive citations and to reproduce diagrams from that indispensable journal.

These secondary sources have been supplemented by information from a series of interviews with leading figures in the UK satellite television business in 1988. I am very grateful to all those who gave me their time, and access to information from their companies. I have respected the desire of most interviewees that their statements remain unattributable.

Readers should be more than usually cautious about the data (and the judgements based on the data) put forward in this study. Two important caveats apply. First, the field is changing very rapidly and information is accordingly highly perishable. Second, much of the information necessary to make well informed judgements is either a matter of commercial confidentiality or simply does not exist. Nonetheless it seems to me that an attempt to assimilate and organize the best information available in the public domain and represent it (so that readers will be better able to make informed judgements without extensive and tedious monitoring of the trade press tea leaves), has been worth attempting. The degree to which the attempt has been successful is for the reader to judge. For the inevitable updates that a rapidly changing field demands see *Cable and Satellite Europe* and other trade journals!

I also owe thanks to Sophie Fahlbeck, Suzanne Hasselbach, Ole Hoelseth and Pat Llewellyn, each of whom have generously shared their knowledge of satellite television with me. To Jay Stuart and Marta Wohrle, not just for editing excellent trade journals, but also for their willingness to share their knowledge, to discuss broadcasting policy, and to purge me of at least some silly ideas. To Florabel Campbell Atkinson who did lots of typing and Terry Staples who did lots of copy editing. And to Manuel Alvarado who encouraged me to prepare this study for publication. None of the above mentioned are responsible for errors or omissions though they can all take some credit for whatever merits this study has.

Readers who still want to know more are commended to Ralph Negrine's useful edited collection *Satellite broadcasting. The politics and implications of the new media* (Routledge, London, 1988) and to my articles ('The Prognosis for Satellite Television in the UK' in *Space Policy*, vol. 5 no. 1, February 1989, Butterworth, Guildford, and 'The

Language of advantage. Satellite television in Western Europe' in *Media Culture and Society*, vol. 11, No. 3, 1989, Sage, London. Both reprinted in my book, *Television: Policy and Culture*, Unwin Hyman, London, 1990) which have a more analytical emphasis than does this present, data led, study. Much of the research on which this study was based was carried out in the Centre for Communication and Information Studies (CCIS) at the Polytechnic of Central London. It was part of CCIS' research carried out for the Economic and Social Research Council (ESRC) and funded under the ESRC's Programme on Information and Communication Technologies (PICT). I owe Nigel Gardner, Director of PICT, many thanks for commissioning me to write the PICT policy paper which forms the final chapter of this revised edition and for his permission to reprint it here.

Bibliography

W. Bekkers (1987) 'The Dutch Public Broadcasting Services in a Multi Channel Landscape'. *European Broadcasting Review (Programmes, Administration, Law)*, Vol XXXVIII, no 6, pp. 32–38, November, Geneva.

Broadcast, Weekly, London.

M. Brynin (1986) *Television in Five European Countries*, Centre for Television Research, University of Leeds, Mimeo.

BSkyB press releases.

Cable & Satellite Europe London, 531-533 Kings Road, London SW10 0TZ, (monthly).

Cablegram (monthly), Newsletter of the Cable Television Association. London.

Canada/Quebec, (1985) *L'Avenir de la television francophone*, Mimeo, Ministries of Communication of Canada and Quebec.

Canal Plus Magazine and press pack, Canal Plus, Paris.

J. Chaplin (1986) 'Satellite Systems Development in Europe and Opportunities in Information Dissemination'. *European Space Agency Bulletin 47*. August, Paris.

R. Collins. (1986) 'Broadband Black Death Cuts Queues. The Information Society and the UK'. in *Media Culture and Society*. vol. 5, no. 3/4, London.

Communications Week International. Manhasset NY Weekly.

Daily Telegraph. Six days a week. London.

DTI. (Department of Trade and Industry) (1988) *Report on the Potential for Microwave Video .*

Distribution Systems in the UK, by Touche Ross. HMSO London.

Deutsche Bundespost (DBP) Press release 29.1.88.

The Economist. weekly London.

A. Ehrenberg and P Barwise (1982) *How much does UK Television cost?* London Business School London.

M. Epstein (1988) 'The Canal Plus Factor' in *Television Business International*, no. 1, pp. 32–41, February, London.

ESA (European Space Agency) (1986) mimeo. *Race Development Phase. A Scenario Using Satellite Television.*

Financial Times London six days a week.

A. Gahlin & B. Nordstrom (1987) 'Access to and use of video and Foreign TV channels in Sweden'. *Sveriges Radio Audience and Programme Research Newsletter*, no. 3, Stockholm. November.

Home Office (1981) *Direct Broadcasting by Satellite*. HMSO London.

Home Office. (1987) *Subscription Television*. HMSO London.

Intermedia. International Institute of Communication. London. Bimonthly.

Kagan Euromedia Regulation. London. Monthly.

M. Long (1987) *World Satellite Almanach*. H. Sams & Co Indianapolis.

P. Marshall (1987) Intelsat – Serving the Needs of Broadcasters. in *European Broadcasting*. Union Review (*Programmes, Administration, Law*). Vol XXXVIII, no. 3, May, pp. 16–18, Geneva.

Media Perspektiven (monthly) Frankfurt.

R. Negrine (ed.) (1988) *Satellite Broadcasting. The Politics and Implications of the New Media* Routledge London.

New Media Markets (NMM). London (fortnightly).

A. Peacock (Chairman) (1986) *The Report of the Committee on Financing the BBC. (The Peacock Report)* HMSO London.

J. N. Pelton and J Howkins (ed.) (1987) *Satellites International*. Macmillan Basingstoke and Stockton Press New York.

Saatchi & Saatchi Compton (1987) *The Media Landscape Now to 1995*. London.

Satellite TV Finance. London. Fortnightly.

Satellite Times. London. Fortnightly.

Screen Digest. Monthly. London.

P. Sepstrup. (1985) *Commercial Transnational and Neighbour Country TV in Europe: Economic Consequences and Consumer Perspectives*. Institut for Markedsokonomie. Handelshojskolen i Aarhus. Aarhus.

Sunday Times. Weekly. London.

Television Business International (TBI) London (ten times a year).

Variety. Weekly. New York.

What Satellite? London. Monthly.

Presspacks and promotional information from Astra, BSB, Childrens' Channel, Disney Channel, Eutelsat, Lifestyle, MTV, Premiere, Screensport, Sky Channel, Super Channel, Worldnet.

Where tables and illustrations are not specifically acknowledged in the text the source is *Cable & Satellite Europe*.

Media titles available from John Libbey

ACAMEDIA RESEARCH MONOGRAPHS

Satellite Television in Western Europe (revised edition 1992)
Richard Collins
Hardback ISBN 0 86196 203 6

Beyond the Berne Convention
Copyright, Broadcasting and the Single European Market
Vincent Porter
Hardback ISBN 0 86196 267 2

The Media Dilemma:
Freedom and Choice or Concentrated Power?
Gareth Locksley
Hardback ISBN 0 86196 230 3

Nuclear Reactions: A Study in Public Issue Television
John Corner, Kay Richardson and Natalie Fenton
Hardback ISBN 0 86196 251 6

Transnationalization of Television in Western Europe
Preben Sepstrup
Hardback ISBN 0 86196 280 X

The People's Voice: Local Radio and Television in Europe
Nick Jankowski, Ole Prehn and James Stappers
Hardback ISBN 0 86196 322 9

Television and the Gulf War
David E. Morrisson
Hardback ISBN 0 86196 341 5

Contra-Flow in Global News
Oliver Boyd Barrett and Daya Kishan Thussu
Hardback ISBN 0 86196 344 X

CNN World Report: Ted Turner's International News Coup
Don M. Flournoy
Hardback ISBN 0 86196 359 8

Small Nations: Big Neighbour
Roger de la Garde, William Gilsdorf and Ilja Wechselmann
Hardback ISBN 0 86196 343 1

BBC ANNUAL REVIEWS

Annual Review of BBC Broadcasting Research: No XV - 1989
Paperback ISBN 0 86196 209 5

Annual Review of BBC Broadcasting Research: No XVI - 1990
Paperback ISBN 0 86196 265 6

Media titles available from John Libbey

Annual Review of BBC Broadcasting Research: No XV - 1991
Paperback ISBN 0 86196 319 9
Peter Menneer (ed)

BROADCASTING STANDARDS COUNCIL PUBLICATIONS

A Measure of Uncertainty: The Effects of the Mass Media
Guy Cumberbatch and Dennis Howitt
Hardback ISBN 0 86196 231 1

Violence in Television Fiction: Public Opinion and Broadcasting Standards
David Docherty
Paperback ISBN 0 86196 284 2

Survivors and the Media
Ann Shearer
Paperback ISBN 0 86196 332 6

Taste and Decency in Broadcasting
Andrea Millwood Hargrave
Paperback ISBN 0 86196 331 8

A Matter of Manners? – The Limits of Broadcast Language
Edited by Andrea Millwood Hargrave
Paperback ISBN 0 86196 337 7

BROADCASTING RESEARCH UNIT MONOGRAPHS

**Quality in Television –
Programmes, Programme-makers, Systems**
Richard Hoggart (ed)
Paperback ISBN 0 86196 237 0

Keeping Faith? Channel Four and its Audience
David Docherty, David E. Morrison and Michael Tracey
Paperback ISBN 0 86196 158 7

**Invisible Citizens:
British Public Opinion and the Future of Broadcasting**
David E. Morrison
Paperback ISBN 0 86196 111 0

School Television in Use
Diana Moses and Paul Croll
Paperback ISBN 0 86196 308 3

Media titles available from John Libbey

UNIVERSITY OF MANCHESTER BROADCASTING SYMPOSIUM

And Now for the BBC ...
Proceedings of the 22nd Symposium 1991
Nod Miller and Rod Allen (eds)
Paperback ISBN 0 86196 318 0

Published in association with UNESCO

Video World-Wide: An International Study
Manuel Alvarado (ed)
Paperback ISBN 0 86196 143 9

Published in association with
THE ARTS COUNCIL of GREAT BRITAIN

Picture This: Media Representations of Visual Art and Artists
Philip Hayward (ed)
Paperback ISBN 0 86196 126 9

Culture, Technology and Creativity
Philip Hayward (ed)
Paperback ISBN 0 86196 266 4

ITC TELEVISION RESEARCH MONOGRAPHS

Television in Schools
Robin Moss, Christopher Jones and Barrie Gunter
Hardback ISBN 0 86196 314 8

Television: The Public's View
Barrie Gunter and Carmel McLaughlin
Hardback ISBN 0 86196 348 2

The Reactive Viewer
Barrie Gunter and Mallory Wober
Hardback ISBN 0 86196 358 X

REPORTERS SANS FRONTIÈRES

1992 Report
Freedom of the Press Throughout the World
Paperback ISBN 0 86196 369 5

MEDECINS SANS FRONTIERES

Populations in Danger
François Jean
Paperback ISBN 0 86196 392 X

Media titles available from John Libbey

IBA TELEVISION RESEARCH MONOGRAPHS

Teachers and Television:
A History of the IBA's Educational Fellowship Scheme
Josephine Langham
Hardback ISBN 0 86196 264 8

Godwatching: Viewers, Religion and Television
Michael Svennevig, Ian Haldane, Sharon Spiers and Barrie Gunter
Hardback ISBN 0 86196 198 6
Paperback ISBN 0 86196 199 4

Violence on Television: What the Viewers Think
Barrie Gunter and Mallory Wober
Hardback ISBN 0 86196 171 4
Paperback ISBN 0 86196 172 2

Home Video and the Changing Nature of Television Audience
Mark Levy and Barrie Gunter
Hardback ISBN 0 86196 175 7
Paperback ISBN 0 86196 188 9

Patterns of Teletext Use in the UK
Bradley S. Greenberg and Carolyn A. Lin
Hardback ISBN 0 86196 174 9
Paperback ISBN 0 86196 187 0

Attitudes to Broadcasting Over the Years
Barrie Gunter and Michael Svennevig
Hardback ISBN 0 86196 173 0
Paperback ISBN 0 86196 184 6

Television and Sex Role Stereotyping
Barrie Gunter
Hardback ISBN 0 86196 095 5
Paperback ISBN 0 86196 098 X

Television and the Fear of Crime
Barrie Gunter
Hardback ISBN 0 86196 118 8
Paperback ISBN 0 86196 119 6

Behind and in Front of the Screen - Television's Involvement with Family Life
Barrie Gunter and Michael Svennevig
Hardback ISBN 0 86196 123 4
Paperback ISBN 0 86196 124 2